Soilborne Plant Pathogens:

Management of Diseases with Macro- and Microelements

Edited by Arthur W ⸺ ⸺

D0881526

APS PRESS
The American Phytopathological Society
St. Paul, Minnesota

This book has been reproduced directly from
computer-generated copy submitted in final form
to APS Press by the editor of the volume. No editing
or proofreading has been done by the Press.

Library of Congress Catalog Card Number: 89-85115
International Standard Book Number: 0-89054-101-9

Printed in the United States of America

The American Phytopathological Society
3340 Pilot Knob Road
St. Paul, Minnesota 55121, USA

TABLE OF CONTENTS

PREFACE

The genesis for this book occurred in a discussion during the 1984 meeting of the A.P.S. Soil Microbiology and Root Disease Committee when we were exploring topics for special programs to sponsor or develop. I had a special interest in the control of Fusarium wilt by manipulation of plant nutrients and had worked closely with several colleagues who were similarly interested. Very successful programs were developed in Florida for Fusarium wilt control in vegetable and ornamental crops by integrating the use of macro- and microelement nutrition, cultural practices, chemotherapeutants and fumigants. The names of other scientists were proposed who had investigated the management of plant diseases caused by soilborne plant pathogens by the manipulation of nutritional amendments. Subsequently, a list of authors and titles for a book was presented for consideration to the A.P.S. Press Chairman, Dr. George N. Agrios. The potential publication was unique in that no other book on this subject was known to be in print. The A.P.S. Press Committee approved the book in 1987.

A broad range of crops and diseases is covered in this book. The interactions of many nutritional elements, soil pH, and cultural factors are discussed in the chapters on Fusarium wilt of vegetables, cotton, and ornamentals; take-all of wheat and other small grains; Verticillium wilt of cotton, alfalfa, and other crops; *Sclerotium rolfsii*-incited diseases of various crops, especially carrots; the peanut pod rot disease complex; aflatoxin contamination of corn and peanuts; Thielaviopsis root rot of cotton and other crops; and scab of potato. The special significance of Ca is detailed in chapters on Erwinia soft rot of potato, club root of crucifers, diseases caused by *Sclerotium rolfsii*, decay in the peanut pod rot complex, Pythium root rot, aflatoxin contamination in corn and peanuts, and Fusarium wilt of vegetables, ornamentals, and cotton.

The editor acknowledges the valuable contribution made by each of the authors to a subject long neglected by plant pathologists and others involved in the growth and production of healthy plants. I also wish to thank LaVerne C. Barnhill, Biological Scientist, for keeping track of the status of the chapters, the reviews, and the correspondence, and also to give thanks to the chapter reviewers who gave so generously of their time and knowledge, and to Dr. Don Huber who formatted and printed the manuscripts on his laser printer. Special thanks to our office staff for typing correspondence, and to members of the Soil Microbiology and Root Disease Committee who encouraged the original concept for this pioneering venture.

Arthur W. Engelhard
Past Chairman
Soil Microbiology & Root Disease Committee
Gulf Coast Research & Education Center
University of Florida Brandenton, Fl 34203

INTRODUCTION

Don M. Huber
Department of Botany and Plant Pathology
Purdue University
West Lafayette, Indiana 47907

Nutrition, although frequently unrecognized, always has been a primary component of disease control. Early settlers moved onto "fresh" soils as readily available nutrients were depleted and disease severity increased. Crop rotation and fallowing practices made crop production possible by increasing the supply of readily available nutrients and controlling weeds which competed for nutrients and moisture. Both factors markedly contributed to disease reduction and remain standard practices for controlling many pests. Cultural disease control tactics such as crop sequence, organic amendment, liming for pH adjustment, tillage, and irrigation frequently influence disease through nutrient interactions (7,8,11-15). These practices supply nutrients directly or render them more or less soluble through altered microbial activity. The advent of readily available inorganic fertilizers for agriculture has brought about the demise of many diseases through improved plant resistance, disease escape, altered pathogenicity, or microbial interactions influencing these (7,8). The effect of N and P on take-all and Pythium root rot of cereals is especially notable.

Nutrition influences all parts of the disease "pyramid" (1). Although we acknowledge the presence of most mineral "cycles," much is still to be learned of their dynamic interactions in the environment with the plant and pathogen (Fig. 1). If the dynamics of most nutrient cycles are considered, it is not surprising to find that some forms of biological control and many "suppressive soils" (soils where disease incidence and severity remain relatively low even with the introduction of a specific pathogen) (9) are manifestations of microbial activity which influence nutrient availability (10-12).

Nutrition of the plant can be drastically altered by many pathogens (Table 1) and it is frequently difficult to clearly differentiate between the biotic and abiotic factors which

1

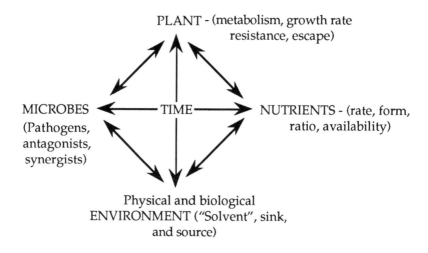

Figure 1. Dynamic interactions influencing the manifestation of disease.

interact to "cause" a nutrient deficiency or excess (7). Thus, through altered uptake, translocation, and distribution, many localized and systemic symptoms of pathogenesis are similar to abiotically-induced nutrient deficiencies or excesses. Many soilborne pathogens may immobilize nutrients such as Mn and Fe outside the plant to create either generalized or localized nutrient deficiencies. Notable in their ability to immobilize nutrients are pathogens such as *Gaeumannomyces graminis* (Sacc.) von Arx & Olivier, *Fusarium oxysporum* Snyder & Hansen, *Verticillium dahliae* Kleb., and *Streptomyces scabies* (Thaxter) Waksman & Henrici which can oxidize Mn^{++} to the non-available Mn^{++++} form in the rhizosphere.

All of the essential mineral elements are reported to influence disease incidence or severity (5,7,8). The effect of mineral nutrients on disease has been determined by: a) observing the effect of mineral amendment (fertilization) on disease severity, b) comparing mineral concentrations in resistant and susceptible cultivars or tissues, c) correlating conditions influencing mineral availability with disease incidence or severity, or d) a combination of all three (8,13). A particular element may reduce some pathogens but increase others, and have an opposite effect with modification of the environment or rate (9). An example is the effect of specific forms of N on diseases of potato (Table 2). Inhibiting nitrification of NH_4-N reduces Verticillium wilt but may increase Rhizoctonia canker. Other cultural modifications can

Table 1. Effects of disease on plant nutrition.

Disease Type	Effect on plant nutrition
Root rots, damping-off, nematodes	Immobilization, solubilization, absorption, distribution
"Masceration" diseases	Distribution ("sinks"), depletion, metabolic efficiency
Vascular wilts	Translocation, distribution
Leaf spots, blights	Metabolic efficiency, distribution
Galls, brooms, overgrowths	Distribution (sinks), metabolic efficiency
Viruses	Sinks, depletion, metabolic efficiency
Fruit and storage rates	Sinks, distribution, reserves

be utilized to reduce Rhizoctonia canker so the NH_4 form of N could be used to suppress the more damaging Verticillium wilt in the production system.

In assessing a nutrient interesting knowledge of the nutrient status of the plant is important. Responses to a particular nutrient may be different when going from deficiency to sufficiency than from sufficiency to excess (N, P, K, Mn, Ca, S). Metabolic systems may respond differently depending on the form of a nutrient (e.g. NH_4-N vs. NO_3-N) or availability (e.g. Mn or S, oxidized versus reduced). Nutrient availability may vary depending on environmental conditions, the previous crop, microbial activity in the rhizosphere, or ratio with other nutrients (N, K, Mn, S, Fe, Zn, Cu) (5,7-9,11-13). Time of application (stage of crop growth) also may influence the disease response to a nutrient. Thus, fall application of N to winter wheat had no deleterious effect on eyespot (*Pseudocercosporella herpotrichoides* (Fron.) Shaw) (Fig. 2) or sharp eyespot (*Rhizoctonia solani* Kuehn), but early spring N top-dressing predisposed plants to these diseases (Table 3) (6).

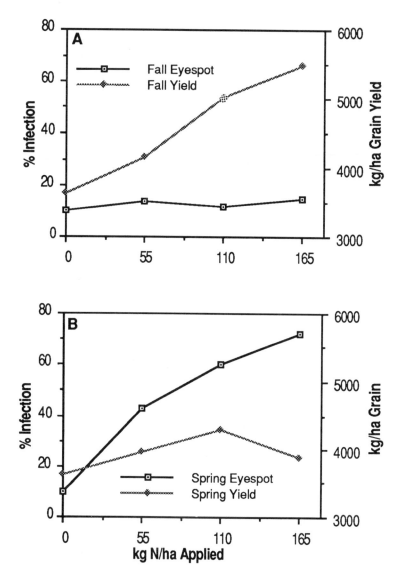

Figure 2. Effect of time of fertilization on foot rot (eyespot) and yield of winter wheat. Yields were higher and disease lower with fall applied (A) than spring applied (B) N (3).

Table 2. Effect of inhibiting nitrification on disease and yield of potato (Watson, R.D. and Huber, D.M. unpublished).

Nitrogen Source	Verticillium Wilt[1]	Rhizoctonia canker[2]	Yield (kg ha^{-1})	% No. 1
$(NH_4)_2SO_4$	3.9 b[3]	6.2 b	6110 b	69 b
$Ca(NO_3)_2$	9.6 a	4.8 a	3598 a	57 a

[1] Indexed on a scale of 0-10 where 0 = no infection; 10 = dead.
[2] Indexed on a scale of 0-10 where 0 = no infection; 10 = complete stem girdling and death.
[3] Means not followed by the same letter are significantly different at 0.05% probability.

Physiological processes influenced by specific nutrients which condition disease reactions are not generally known. Some of these processes are probably general mechanisms as indicated in the case of Mn which has a critical function in phenol metabolism and lignification, as well as N and carbohydrate metabolism and photosynthesis (2-4). Although a general system may be involved, it may act more specifically in a particular host-pathogen interaction. An example is the contrasting effect of N forms on Verticillium wilt of potato relative to the concentration of phenolic-compounds in the vascular elements which are inhibitory to the pathogen (Fig. 3). Nitrate-N inhibits phenol metabolism (4) and hastens disease, while NH_4-N delays symptom expression. Since mineral elements are the

Table 3. Effect of time of fertilization on sharp eyespot of winter wheat.

N Applied	Lodging	Disease Severity*	Yield
	%		kg/ha
Fall	3	3.1	3036
Spring	73	4.2	2640

* Tillers with sharp eyespot lesions: 0 = no disease 1 = < 10%; 2 = 11-20%; 3 = 24-30%; 4 = 31-40%; 5 = > 40%.

activators, inhibitors, and regulators of metabolic systems, it is doubtful that an understanding of most host-pathogen interactions will be gained until the associated nutrient relationships are elucidated.

Disease control through nutrient manipulation may be accomplished by modifying: 1) nutrient availability through amendment (fertilization) or by changing the physical (pH, moisture, aeration) and/or biological (root exudates, seed bacterization with plant growth promoting rhizobacteria) environment, or 2) nutrient uptake by improved cultivar efficiency (4,5,13) or tolerance (13). Breeding nutrient efficient or tolerant crops and establishing predictable cultivar requirements will improve production efficiency through improved disease control and physiological function.

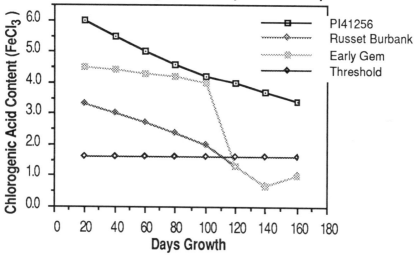

Figure 3. Effect of cultivar, stage of growth, and form of nitrogen on chlorogenic acid in the vascular elements of potato and resistance to Verticillium wilt (Watson, R.D. and Huber D.M., unpublished). Threshold is the level of phenol required to inhibit extracellular enzymes of Verticillium albo-atrum. The sharp drop in phenol content below the threshold concentration in Early Gem and Russet Burbank around 120 days corresponds with tuber initiation or "flowering" and early onset of symptoms. The Verticillium resistant cultivar PI41259 maintains a phenol concentration above the threshold for enzyme inhibition until maturity.

The flexibility in most disease-nutrient interactions permits the much broader utilization of this cultural control in reducing disease severity than is presently practiced. It is clear that the severity of most diseases can be reduced and the chemical, biological, or genetic control of many plant pathogens enhanced by proper nutrition. The following chapters contain numerous examples of how nutrition influences 1) the growth, virulence, reproduction/replication, and survival of pathogens; 2) predisposition, escape, tolerance, and resistance of plants; and 3) root exudates and microbial interactions important in biological control.

Nutrient manipulation through amendment or modification of the soil environment is an important cultural control for plant disease and an integral component of production agriculture. It is hoped that the information presented in the following chapters will stimulate the research needed to more effectively use this cultural control of diseases and improve crop production efficiency.

LITERATURE CITED

1. Bateman, D.F. 1978. The dynamic nature of disease. Pages 53-85 in: Plant Pathology-An Advanced Treatise. Vol. 3. J.G. Horsfall and E.B. Cowling, eds. Academic Press, NY.
2. Burnell, J.N. 1988. The biochemistry of manganese in plants. Pages 139-154 in: Manganese in Soils and Plants. R.D. Graham, R.J. Hannam, and N.C. Uren, eds. Kluwer Academic Publishers, Dordrecht, Netherlands.
3. Campbell, L.C., and Nable, R.O. 1988. Physiological functions of manganese in plants. Pages 139-154 in: Manganese in Soils and Plants. R.D. Graham, R.J. Hannam, and N.C. Uren, eds. Kluwer Academic Publishers, Dordrecht, Netherlands.
4. Graham, R.D. 1988. Genotypic differences in tolerance to Manganese deficiency. Pages 261-292 in: Manganese in Soils and Plants. R.D. Graham, R.J. Hannam, and N.C. Uren, eds Kluwer Academic Publishers, Dordrecht, Netherlands.
5. Graham, R.D., and Webb, M.J. 1989. Micronutrients and Disease Resistance and Tolerance in Plants. in: Micronutrients in Agriculture, 2nd Edition. R.M. Welch (ed). Am. Soc. Agron., Madison, WI.
6. Huber, D.M. 1976. Plant Disease. Pages 325-327 in: McGraw-Hill Yearbook of Science and Technology. McGraw-Hill, New York, NY.
7. Huber, D.M. 1978. Disturbed mineral nutrition. Pages 163-181.in: Plant Pathology-An Advanced Treatise, Vol. 3. J.G.

Horsfall and E.B. Cowling, eds. Academic Press, NY.

8. Huber, D.M. 1980. The role of nutrition in defense. Pages 381-406 in: Plant Pathology-An Advanced Treatise, Vol. 5. J.G. Horsfall and E.B. Cowling, eds. Academic Press, N.Y.

9. Huber, D.M., and Arny, D.C. 1985. Interactions of potassium with plant disease. Pages 467-488 in: Potassium in Agriculture. D. Munson, ed. Am. Soc. Agron., Madison, WI.

10. Huber, D.M., and Schneider, R.W. 1982. The description and occurrence of suppressive soils. Pages 1-7 in: Suppressive Soils and Plant Disease. R.W. Schneider ed. Am. Phytopathol. Soc., St. Paul, Mn.

11. Huber, D.M., and Watson. R.D. 1970. Effect of organic amendment on soilborne plant-pathogens. Phytopathology 60:22-26.

12. Huber, D.M., and Watson, R.D. 1974. Nitrogen form and plant disease. Annul. Rev. Phytopathol. 12:139-165.

13. Huber, D.M., and Wilhelm, N.S. 1988. The role of Manganese in resistance to plant diseases. Pages 155-173 in: Manganese in Soils and Plants. R.D. Graham, R.J. Hannam, and N.C. Uren, eds. Kluwer Academic Publishers, Dordrecht, Netherlands.

14. McNew, G.L. The effects of soil fertility. Pages 100-114 in: Plant Diseases, the Yearbook of Agriculture, 1953. U.S. Dept. Agr., Washington, D.C.

15. Office of Technology Assessment. 1979. Pest Management Strategies in Crop Protection. Vol. I, II. Congress of the United States, Washington, D.C.

HISTORICAL HIGHLIGHTS AND PROSPECTS FOR THE FUTURE

Arthur W. Engelhard
Gulf Coast Research & Education Center
IFAS, University of Florida
Bradenton, FL 34203

The application of fertilizer is a universal practice in the production of commercial crops. Macro- and microelements long have been recognized as being associated with changes in the level of disease and size and/or yields of plants. The physiological roles associated with the elements may be more or less understood but direct linkage of biochemical and physiological factors such as enzymes, cell wall and membrane permeability, competing ions, amino acids, phenols, alkaloids, phytoalexins, electron transport, etc. with disease is rather incomplete. The complex balance of interdependent functions of the host are changed and influenced by chemical elements which in turn influence the reaction to disease. At times, however, there is a simple and direct relationship in physiological plant pathology.

Much the same situation exists for the physiology of soil microflora, including the plant pathogens. The soil environment and severity of diseases in plants are changed by macro and microelement additions. Nitrogen stimulates the soil microflora so that competition for nutrients and space decreases the relative number(s) of propagules of the pathogen(s). Antagonists may be stimulated to produce more antibiotics. Calcium and NO_3-N raise the soil pH, decrease the virulence of *Fusarium oxysporum* and reduce the level of Fusarium wilt whereas NH_4-N decreases the pH of the soil and rhizosphere and restricts the growth of the take-all pathogen, *Gaeumannomyces graminis* (Sacc.) von Arx and Oliver (*Ophiobolus graminis*). Nitrogen, P, K and Ca reduce damping-off. Excess N (imbalance) can cause thin cell walls allowing fungi to penetrate more readily while balanced nutrition can increase vigor and reduce damping-off. The level of Fusarium wilt increases while Verticillium wilt decreases when the soil pH (or Ca) is low. Phosphorus and K can promote root growth and

9

reduce seedling disease and root rots. The effects (above) of nutrients in changing the level of disease were extracted from Huber (11-13), Jones and Woltz (17,20-23), Graham (10), Huber and Arny (14), McNew (28) and Sadasivan (32).

SOME CLASSIC EXAMPLES

It has been known for nearly 100 years that amending the soil with $CaCO_3$ would provide a significant measure of disease control of club root of crucifers. Woronin (50) in Russia described club root of cabbage and other crucifers (*Plasmodiphora brassicae* Wor.) in 1878. Club root was more severe in crops grown on acid soils and extensive research was conducted to control the disease with lime. Cunningham (2) in 1914, increased yield 3.5 times by adding $CaCO_3$, but it was Campbell, Greathead, Myers and deBoer (1) in 1985 who got up to 3 years field control of club root of broccoli in the Salinas Valley of California with a single application of $CaCO_3$. They postulated "that successful control by liming depended on the interaction between pH and extractable Ca plus Mg". This represents a classical success story of plant pathology that covered nearly 100 years.

Thaxter published in 1891 (42) that corky scab of potato was incited by *Actinomyces scabies* (Thax.) Gussow. Martin in 1922 and 1924 (26,27) indicated that scab was less severe on soils with an acid reaction than an alkaline reaction, indicating the need for soil amendments with S compounds or acidic fertilizers to aid in disease control.

The take-all disease of wheat according to McNew in 1953 (28) has a long history of exacting severe losses in the U.S. and in other locations. Extensive research programs by many investigators studying such factors as N level, N source, timing of fertilizer applications, P, K, chlorides, sulfates and biorationals (13,29,44) have resulted in the development of disease control components which, when integrated in the cultural program, maximized disease control and yield.

Fusarium wilt is one of the most serious diseases worldwide. It is incited by specific formae speciales of *Fusarium oxysporum* that infect many cultivars and species of plants. In 1920, Edgerton and Moreland (3) in Louisiana demonstrated that adding large amounts of lime reduced the level of wilt in tomato. Many papers relating to pH, Ca, N and form of N, P, K and other soil amendments are in the literature. Many researchers studied this important disease but positive results were not integrated to maximize disease control until Jones and Woltz (16-23) and Woltz and Jones (46-49) in Florida expanded the *Fusarium* research on tomatoes in the 1960's and 1970's. They studied NO_3-N vs NH_4-N, microelements, P, $CaCO_3$, $Ca(OH)_2$ and $CaSO_4$. They found the pathogen could be starved and its virulence reduced by macro- and microelement manipulation. They combined their positive leads into a cultural system which consistently gave high levels of disease control and increased yields in tomatoes, cucumbers and watermelons in the sandy soils of southwest Florida. Engelhard (4-6), Engelhard and Woltz (7-9) and Woltz and Engelhard (45) continued in the '70s and developed the integrated nitrate-lime-chemotherapy wilt control system that provided complete disease suppression in floral crops such as chrysanthemum and aster where total control was needed to produce high quality flowers. They postulated that the integrated nitrate-lime-chemotherapy system would be efficacious on other crops. The system worked satisfactorily on tomatoes in Hungary for Sarhan and Kiraly (33), on greenhouse chrysanthemums in Maryland (25) and commercial chrysanthemums in Florida (31) and North Carolina (38,39). The scenario for Fusarium Wilt is not complete for there also are new leads such as siderophores (51), *Pseudomonas putida* (34), *Alcaligenes* sp. (51), special soil amendments (40), hardwood bark amendments (41), soil suppressive factors (34,35), Fe and Fe chelating compounds (34,35,37), and chlorides (36). These are some of the new leads that can continue to expand the knowledge and understanding of the biology of *Fusarium* and improve control. If the positive disease control factors of the present research and the wealth of information available in the literature are integrated, other successful systems for Fusarium wilt control may be developed for use on the many wilt susceptible crops growing in the varied climates and soil environments of the world.

AVAILABLE LITERATURE

A large volume of literature is available on disease control effects provided by macro- and microelement amendments. Huber and Watson in 1974 (15) in "Nitrogen Form and Plant Disease" reviewed and discussed the effects of N and/or N form on seedling disease, root rots, cortical diseases, vascular wilts, foliar diseases and others. They summarized work from the 259 references in four tables in which they list crops, diseases and citations. McNew (28) in the 1953 USDA Yearbook of Agriculture discussed effects of fertilizers on soilborne diseases and their control. He reviewed briefly specific diseases such as take-all of wheat, Texas root rot, Fusarium wilt of cotton, clubroot of crucifers and common scab of potato. Many other diseases were mentioned, as well as how macro- and microelements effect host physiology and disease. Huber and Arny (14) in "Interactions of Potassium with Plant Disease" summarized in three tables the effect of K (positive, negative, neutral) on specific diseases. They listed 267 references in the bibliography.

The Potash and Phosphate Institute is dedicated to research and education and celebrated its 50th anniversary in 1985. It is a source of information on the use of K and P in the production of plants and the effects on plant disease. The Institute promotes a systems approach to crop production; disease control is one of the factors in the system (43).

Leath and Ratcliffe (24) described plant nutrition and diseases in forage crops production. They indicated that fertilizers affect pathogens in the soil and on the host, and also can effect the pathogenicity of an organism. Presley and Bird (30) reviewed the effect of P on the reduction of disease susceptibility of cotton.

In 1983, Graham, in Australia in "Effects of Nutrient Stress on Susceptibility of Plants to Disease with Particular Reference to the Trace Elements" (10), discussed under the heading "Macroelements," the effect of six essential elements on disease; and under "Micronutrients," seven essential elements and 15 others as having been reported to influence a host-parasite relationship. He gives 305 literature citations.

Another review by Huber (13) entitled, "The Use of Fertilizers and Organic Amendments in the Control of Plant Disease" contains a wealth of information. He indicated how the severity of 157 diseases was affected by N in Table 1. In Table 2, a similar listing is given for NO_3- and NH_4-forms of N. The effects of P, K, Ca and Mg are given in Tables 3, 4, 5 and 6, respectively. Tables for S, Na, Mn, Fe, Zn, B, Cu, Si and other elements also are presented.

A literature search of the CAB ABSTRACTS data base utilizing the DIALOG Information Retrieval Service and using some key words; soilborne disease, macroelements, microelements, soil fungi, *Fusarium*, *Pythium*, and *Phytophthora*, yielded 1500 citations published during the past 14 years.

THE FUTURE

Obviously a virtual flood of literature is available regarding the effects of macro- and microelement soil amendments on the level of soilborne disease in plants. What is lacking is the correlation of the positive factors into integrated production systems. The biggest problem now is how to organize and comprehend the mountain of available and often conflicting data. We have entered an era in which computer-aided analyses and other sophisticated tools are needed to integrate information and develop systems approaches to growing healthy, productive plants. One of the most rewarding approaches for successful reduction of soilborne diseases is the proper selection and utilization of macro- and microelements. Since virtually all commercially produced crops in the developed world are fertilized, it is extremely important to select macro- and microelements that decrease disease. This is an important and viable alternative or supplement to the use of pesticides which usually only give partial disease control.

LITERATURE CITED

1. Campbell, R.N., Greathead, A.S., Myers, D.F., and de Boer, G.J. 1985. Factors related to control of clubroot of crucifers in the Salinas Valley of California. Phytopathology 75:665-670.
2. Cunningham, G.E. 1914. Studies on clubroot. II. Disease resistance of crucifers; methods of combating clubroot. Vt. Agric. Exp. Stn. Bull. 185:65-96.
3. Edgerton, C.W. and Moreland, C.C. 1920. Tomato wilt. La. Agric. Exp. Stn. Bull. 174:3-54.
4. Engelhard, A.W. 1974. Effect of pH and nitrogen source on the effectiveness of systemic fungicides in controlling Fusarium wilt of chrysanthemums. Proc. Am. Phytopathol. Soc.1:121 (Abstr.).
5. Engelhard, A.W. 1975. Aster Fusarium wilt: complete symptom control with an integrated fungicide-NO3-pH control system. Proc. Am. Phytopathol. Soc. 2:62 (Abstr.).
6. Engelhard, A.W. 1979. Control of Fusarium wilt of carnation with an integrated nitrate-nitrogen and systemic fungicide control program. Phytopathology 69:1027 (Abstr.) & IX International Congress of Plant Protection (Abstr. 557).
7. Engelhard, A.W. and Woltz, S.S. 1973. Fusarium wilt of chrysanthemum: Complete control of symptoms with an integrated fungicide-lime- nitrogen regime. Phytopathology 63:1256-1259.
8. Engelhard, A.W. and Woltz, S.S. 1973. Fusarium wilt of chrysanthemum: A new cultural chemical control method. Fla. Flower Grower 10(9):1-3.
9. Engelhard, A.W. and Woltz, S.S. 1978. Effect of temperature, nitrogen source, lime and benomyl treatments on Fusarium wilt of

chrysanthemum, aster and gladiolus. Page 375 in: Third Int. Congress of Plant Pathology, Abstracts of Papers (Abstr.).

10. Graham, R.D. 1983. Effects of nutrient stress on susceptibility of plants to disease with particular reference to the trace elements. Pages 221-276 in: Advances in Botanical Res. Vol. 10. H.W. Woolhouse, ed. Academic Press, London, UK.

11. Huber, D.M. 1980. The role of mineral nutrition in defense. Pages 381-406 in: Plant Disease: An Advanced Treatise, Vol. 5. J.G. Horsfall and E.B. Cowling, eds. Academic Press, New York.

12. Huber, D.M. 1981. The role of nutrients and chemicals. Pages 317-341 in: The Biology and Control of Take-all. PJ. Shipton and M. Asher, eds., Academic Press, London, UK.

13. Huber, D.M. 1981. The use of fertilizers and organic amendments in the control of plant disease. Pages 357-394 in: CRC Handbook of Pest Management in Agriculture. Vol. 1. D. Pimental, ed. CRC Press, Boca Raton, FL.

14. Huber, D.M. and Arny, D.C. 1985. Interactions of potassium with plant disease. Pages 467-488 in: Potassium in Agriculture. R.D. Munson, ed. Am. Soc. Agron., Madison, WI.

15. Huber, D.M. and Watson, R.D. 1974. Nitrogen form and plant disease. Annu. Rev. Phytopathol. 12: 139-165.

16. Jones, J.P. and Woltz, S.S. 1967. Fusarium wilt (race 2) of tomato: Effect of lime and micronutrient soil amendments on disease development. Plant Dis. Rep. 51:645-648.

17. Jones, J.P. and Woltz, S.S. 1968. Field control of Fusarium wilt (race 2) of tomato by liming and stake disinfestation. Proc. Fla. State Hort. Soc. 81:187-191.

18 Jones, J.P. and Woltz, S.S. 1969. Fusarium wilt (race 2) of tomato: calcium, pH and micronutrient effects on disease development. Plant Dis. Rep. 53:276-279.

19. Jones, J.P. and Woltz, S.S. 1970. Fusarium wilt of tomato: interaction of soil liming and micronutrients on disease development. Phytopathology 60: 812813.

20. Jones, J.P. and Woltz, S.S. 1972. Effect of soil pH and micronutrient amendments on Verticillium and Fusarium wilt of tomato. Plant Dis. Rep. 56: 151-153.

21. Jones, J.P. and Woltz, S.S. 1975. Effect of liming and nitrogen source on Fusarium wilt of cucumber and watermelon. Proc. Fla. State Hort. Soc. 88:200-203.

22. Jones, J.P. and Woltz, S.S. 1981. Fusarium- incited diseases of tomato and potato and their control. Pages 157-168 in: Fusarium: Diseases, Biology and Taxonomy. P.E. Nelson, T.A.Toussoun and R.J. Cook, eds. Penn. State Univ. Press, University Park, PA.

23. Jones, J.P. and Woltz, S.S. 1983. Cultural control of Fusarium wilt race 3 of tomato. Proc. Fla. State Hort. Soc. 96:82-83.
24. Leath, K.T. and Ratcliffe, R.H. 1974. The effect of fertilization on disease and insect resistance. Pages 481-503 in: Forage Fertilization. D.A. Mays, ed. A.S.A., C.S.S.A., S.S.S.A., Madison, WI.
25. Locke, J.C., Marois, J.J., and Papavizas, G.C. 1985. Biological control of Fusarium wilt of greenhouse-grown chrysanthemums. Plant Dis. 69:167-169.
26. Martin, W.H. 1922. Potato scab and methods for its control. N.J. Agric. Exp. Stn. Circ. 131:1- 12.
27. Martin, W.H. 1924. Influence of soil moisture and acidity on the development of potato scab. Soil Science 16:69-73.
28. McNew, G.L. 1953. The effects of soil fertility in plant diseases. Pages 100-114 in: Plant Diseases. U.S. Dept. Agric. Yearbook, Washington, D.C.
29. Powelson, R.L. and Jackson, T.L. 1978. Suppression of take-all (Gacumannomyces graminis) root rot of wheat with fall applied chloride fertilizers. Pages 175-182 in: Proc. 28th Annu. Fert. Conf. Pac. Northwest.
30. Presley, J.T. and Bird, L.S. 1968. Diseases and their control. Pages 347-366 in: Advances in Production and Utilization of Quality Cotton: Principles and Practices. F.C. Elliot, M.Hoover, and W.K. Porter, Jr., eds. Iowa State Univ. Press, Ames, IA.
31. Raju, B.C. 1983. Fusarium wilt of mums. Acta Hort. 152:65-76.
32. Sadasivan, T. S. 1965. Effect of mineral nutrients on soil microorganisms and plant disease. Pages 460-469 in: Ecology of Soilborne Plant Pathogens. K.R. Baker and W.C. Snyder, eds. Univ. of Calif. Press, Berkeley, CA.
33. Sarhan, A.R.T. and Kiraly, Z. 1981. Control of Fusarium wilt of tomato with an integrated nitrate-lime-fungicide regime. Acta Phytopathol. Acad. Scient. Hung. 16(1/2):9-14.
34. Scher, F.M. and Baker, R. 1982. Effect of Pseudomonas putida and a synthetic iron chelator on induction of soil suppressiveness to Fusarium wilt pathogens. Phytopathology 72:1567-1573.
35. Scher, F. M., Dupler, M. and Baker, R. 1984. Effect of synthetic iron chelates on population densities of Fusarium oxysporum and the biological control agent Pseudomonas putida in soil. Can. J. Microbio. 30:1271-1275.
36. Schneider, R.W. 1985. Suppression of Fusarium yellows of celery with potassium, chloride, and nitrate. Phytopathology 75:40-48.

37. Simeoni, L.A., Lindsay, W.L. and Baler, R. 1987. Critical iron level associated with biological control of Fusarium wilt. Phytopathology 77:1057- 1061.
38. Strider, D.L. 1985. Fusarium wilt of chrysanthemum: cultivar susceptibility and chemical control. Plant Dis. 69:564-568.
39. Strider, D.L. and Jones, R.K. 1983. Fusarium wilt of pot mums. N.C. Flower Grower Bull. 27(3):10-11.
40. Sun, S.K. and Huang, J.W. 1985. Formulated soil amendment for controlling Fusarium wilt and other soilborne diseases. Plant Dis. 69:917-920.
41. Trillas-Gay, M.I., Hoitink, H.A.J., and Madden, L.V. 1986. Nature of suppression of Fusarium wilt of radish in a container medium amended with composted hardwood bark. Plant Dis. 70:1023-1027.
42. Thaxter, R. 1891. The potato scab. Conn. (State) Stn. Rep. 14:81-95.
43. Usherwood, N.R. 1980. The effects ofpotassium on plant disease. Pages 151-164 in: Potassium for Agriculture, A Situation Analysis. Potash and Phosphate Institute, Atlanta, GA.
44. Weller, D.M. and Cook, R.J. 1983. Suppression of take-all of wheat seed-treatment with fluorescent pseudomonads. Phytopathology 73:463- 469.
45. Woltz, S.S. and Engelhard, A.W. 1973. Fusarium wilt of chrysanthemum: Effect of nitrogen source and lime on disease development. Phytopathology 63:155-157.
46. Woltz, S.S. and Jones, J.P. 1968. Micronutrient effects on the in vitro growth and pathogenicity of *Fusarium oxysporum* f. sp. *lycopersici*. Phytopathology 58:336-338.
47. Woltz, S.S. and Jones, J.P. 1973. Interactions in source of nitrogen fertilizer and liming procedure in the control of Fusarium wilt of tomato. HortScience 8:137-138.
48. Woltz, S.S. and Jones, J.P. 1973. Tomato Fusarium wilt control by adjustments in soil fertility. Proc. Fla. State Hort. Soc. 86: 157-159.
49. Woltz, S.S. and Jones, J.P. 1981. Nutritional requirements of *Fusarium oxysporum*: Basis for a disease control system. Pages 340-349 in: Diseases, Biology and Taxonomy. P.E. Nelson, T. A. Toussoun, and R. J. Cook, eds. Penn. State Univ. Press, University Park, PA.
50. Woronin, M. 1878. *Plasmodiophora brassicae*: The cause of cabbage hernia. Phytopathol. Classic No. 4. Am. Phytopathol. Soc., St. Paul, MN. 32 pp.

51. Yuen, G.Y. and Schroth, M.N. 1986. Inhibition of *Fusarium oxysporum* f. sp. *dianthi* by iron competition with an *Alcaligenes* sp. Phytopathology 76:171-176.

MANAGEMENT OF FUSARIUM WILT OF VEGETABLES AND ORNAMENTALS BY MACRO- AND MICROELEMENT NUTRITION

John P. Jones, Arthur W. Engelhard and S. S. Woltz
Gulf Coast Research & Education Center
5007 60th Street East
Bradenton, FL 34203

Fusarium wilt (*Fusarium oxysporum* Schlecht. f. sp. *lycopersici* (Sacc.) Snyder and Hansen) of tomato (*Lycopersicon esculentum* Mill.) was first reported in the United States in 1899 by E. F. Smith (59) who said that the disease had put an end to growing tomatoes for northern markets in certain areas of Florida. By 1920, the disease had become widespread and destructive throughout the southeast, midwest, and middle Atlantic states (69).

The first research efforts on Fusarium wilt of tomato initiated circa 1910 by Norton (46,47) and Essary (20,21) and concentrated on the development of tolerant varieties. Nearly concurrently, Wollenweber (71) and Sherbakoff (56) started their studies on the morphology and physiology of the genus *Fusarium*. A third phase of research was begun around 1915 by Edgerton and Moreland (12,13) who investigated the effect of nutrition on disease development.

Fusarium wilt of chrysanthemum (*Endranthema grandiflora* Tzvelev.) was recognized as a disease in 1963 and published in 1964 by Armstrong and Armstrong (3). They demonstrated that *F. oxysporum* Schlecht. f. sp. *tracheiphilum* Armstrong & Armstrong race 1 caused wilt of the chrysanthemum cultivar Encore and certain cultivars of cowpea and soybean. In 1970, Armstrong et al. (4) determined that *F. oxysporum* Schlect. f. sp. *chrysanthemi* (Litt., Armstrong & Armstrong) (Foc) was the cause of wilt in the cultivars Encore and Yellow Delaware. The latter forma specialis commonly is found affecting chrysanthemum.

Research on the effects of lime (pH), N sources, and systemic fungicides on the control of Fusarium wilt of chrysanthemum was first pursued by Engelhard and Woltz in the early 1970's (17,18,72).

This review places special emphasis on investigations and concepts which led to the Florida system of Fusarium wilt control on vegetable crops and the lime (pH):nitrate-nitrogen:chemotherapy system on ornamental crops. Crop emphasis in Florida was on tomato and chrysanthemum because of the importance of these two crops, but research on watermelon (*Citrullus vulgaris* Schrad.), cucumber (*Cucumis sativus* L.), muskmelon (*Cucumis melo* L.), King Aster (*Callistephus chinensis* L.), carnation (*Dianthus caryophyllus* L.) and other crops are included in this review.

LIME, CALCIUM, pH

Edgerton and Moreland (12,13) reported as early as 1913 that "lime added to the soil in large amounts, such as ten tons to the acre, hinders the development of the [tomato] wilt". Their finding was confirmed by numerous workers including Sherwood (57) in 1923, Scott (55) in 1924, and Fisher (23,24) in 1935. Throughout the years the association of higher pH and/or Ca with less wilt was shown by many other investigators on numerous crops (Table 1).

The severity of Fusarium wilt of tomato decreased with an increase in calcium (Ca) concentration from 5 to 500 ppm in a nutrient solution experiment of Edgington and Walker (14). Corden (8) and Standaert et al. (61) also found that a Ca deficiency encouraged tomato wilt development and Corden (8) suggested that the increased pathogenesis resulted because there was little Ca to inhibit the activity of the polygalacturonase produced by *Fusarium*.

Jones and Woltz (31,32) found in greenhouse and field experiments that amending *F. oxysporum* f. sp. *lycopersici* infested soil with calcium sulfate ($CaSO_4$- gypsum) did not increase the soil pH and did not reduce the occurrence of tomato Fusarium wilt. In contrast, the Ca content of tissue of plants grown in the gypsum- amended soil was as great as that of plants grown in soil amended with hydrated lime ($Ca(OH)_2$) which did reduce the occurrence of wilt. They concluded that high tissue Ca was not a major factor in controlling Fusarium wilt.

NITROGEN

Clayton (6) in 1923 started investigations to determine the effect of nitrate-nitrogen (NO_3-N) on the severity of Fusarium wilt of tomato. This work was followed by that of Ahmet (1) in 1923 and of Cook (7) in 1937. These workers, in general, compared NO_3-N to no N treatments and obtained inconclusive results. Fisher (23,24) in 1935 probably was

Table 1. Association of high pH or Ca with wilt reduction.

Crop	Investigators	Reference(s)
Tomato	Foster & Walker, 1947	25
(*Lycopersicon*)	Edgington & Walker, 1958	14
	Corden, 1965	8
	Jones & Woltz, 1967,1968,	30,31,32,33
	1969,1970,1972,1981,1983	34,36,37
	Jones & Overman, 1971	29
	Woltz & Jones, 1973, 1984	75,76,77,79
	Standaert et al., 1978	62
	Sarhan & Kiraly, 1981	52
Chrysanthemum	Engelhard & Woltz,1972,	17,18,19
(*Dendranthem*)	1973,1978	
	Woltz & Engelhard, 1973	72
	Raju, 1983	49
	Locke et al., 1985	40
Carnation	Engelhard, 1979	16
(*Dianthus*)	Pergola et al., 1979	48
Cotton	Albert, 1946	2
(*Gossypium*)		
Cucumber	Jones & Woltz, 1975	35
(*Cucumis*)		
Flax	Keane & Sackston, 1970	38
(*Linum*)		
Gladiolus	Woltz & Magie, 1975	80
(*Gladiolus*)	Engelhard & Woltz, 1978	19
King Aster	Engelhard, 1975	15
(*Callistephus*)	Engelhard & Woltz, 1978	19
Muskmelon	Stoddard, 1947	63
(*Cucumis*)		
Radish	Sun & Huang, 1985	65
(*Raphanus*)	Huang et al., 1986	28
Strawberry	Yoshino & Hashimoto, 1978	81
(*Fragaria*)		
Watermelon	Everett & Blasquez, 1967	22
(*Citrullus*)	Jones & Woltz, 1975	35
	Sun & Huang, 1985	65

the first to demonstrate that an increase in the ratio of ammonia-nitrogen (NH_4-N) to NO_3-N increased disease severity. However, he attributed the increased severity to an increase in total N.

Much of the research with N source was combined with pH studies. Albert (2) in 1946 reported that cotton (*Gossypium hirsutum* L.) plants grown in $Ca(NO_3)_2$ at high pH had less wilt (*F. oxysporum* Schlecht. f. sp. *vasinfectum* Atk.) Snyder and Hansen) injury than those grown in any other N containing solution. Jones and Woltz (30-35) and Woltz and Jones (74-77) greatly expanded the early work with N source and lime amendments on tomato. They consistently demonstrated that liming sandy soils to a high pH (6.5-7.5) significantly decreased the incidence and severity of Fusarium wilt of tomato as Scott (55) and other workers (23,24,57) had reported. They also demonstrated that NO_3-N decreased disease development compared to NH_4-N and that the beneficial effects of high soil pH could be overcome by the use of NH_4-N (76,77). They further demonstrated that effects of high soil pH and NO_3-N were additive so that the combination of the two almost invariably resulted in even better wilt control. Methods were developed by Jones, Woltz, and Engelhard in Florida which resulted in consistent control in the field and greenhouse of Fusarium wilt of tomato (30-35,75-77) and other vegetables (35), and of chrysanthemum (17-19,72) and other ornamentals (15,16,19). Jones and Woltz (35) also showed that gross yields were increased and wilt development decreased when NO_3-N, as compared to NH_4-N, was combined with high pH on watermelons and cucumbers. Stoddard (63) also demonstrated that less Fusarium wilt (*F. oxysporum* Schlecht. f. sp. *niveum* (Leach and Currence) Snyder and Hansen) occurred on muskmelon when using a NO_3-N rather than NH_4-N source at pH 6.0. Schneider (54) found more wilt with NH_4-N than with NO_3- N in celery (*Apium graveolens* L. var. *dulce*) wilt (*F. oxysporum* Schlecht f. sp. *apii* (R. Nelson and Sherb.) Snyder and Hansen. Trillas-Gay et al. (67) found significantly reduced disease levels at the highest concentration of $Ca(NO_3)_2$ whereas NH_4NO_2 had no significant effect on Fusarium wilt (*F. oxysporum* Schlecht. f. sp. *raphani* Kendr. and Snyder) of radish (*Raphanus sativus* L.). Engelhard (15,16), Engelhard and Woltz (17-19), and Woltz and Engelhard (72) consistently found less wilt in chrysanthemums, King Aster and carnation with NO_3-N compared to NH_4-N. Locke (40) and Raju (49) reported the same on chrysanthemum. Dick and Tisdale (10) found that N alone and or in certain combinations with K reduced wilt and increased yield of cotton. However, the N source was not given. Morgan and Timmer (45) reported that Fusarium wilt (*F. oxysporum* Schlecht.) severity of Mexican limes (*Citrus aurantifolia* (Christm.) Swingle) and the numbers of propagules were reduced (and the pH higher) in media fertilized with NO_3-N. Spiegel and Netzer (60) in Israel demonstrated that NH_4-N fertilized muskmelon plants showed 30% more Fusarium wilt compared to NO_3-N fertilized plants. Keim and Humphrey (39) reported excellent control of Fusarium wilt (*F. oxysporum* Schlecht.) of Hebe (*Hebe* spp.) in California by using NO_3-N compared to NH_4-N.

Walker and Foster (70) reported that disease development in the highly susceptible cv. 'Bonny Best' tomato variety in nutrient cultures using all NO_3-N increased in severity in the following order: low N, high potassium (K), normal N + K, high N, and low K. This indicated that high N or low K favored disease development, whereas low N or high K retarded it. In contrast, Foster and Walker (25) found that tomato plants preconditioned with low N or high K prior to inoculation were more susceptible than those preconditioned with high N or low K. They also showed that a high pH inhibited wilt development.

Bloom and Walter (5) applied urea-N to tomato foliage weekly for 30 days before inoculation and found a slight increase in disease severity with an increase in concentration.

Loffler et al. in 1986 (41,42) showed that NO_3-N inhibited chlamydospore formation of *F. oxysporum* Schlecht., whereas urea or NH_4Cl reduced chlamydospore formation and lysis. They concluded that NO_3-N but not NH_4-N was involved.

High levels of N, regardless of source, were found by Woltz and Magie (80) to encourage Fusarium corm rot of gladiolus (*Gladiolus hortulanus* Bailey) caused by *F. oxysporum* Schlecht. f. sp. *gladioli* (Massey) Snyder and Hansen. However, NH_4-N resulted in a greater incidence of Fusarium yellows of the mother plants than 50:50 NH_4NO_3-N.

Hopkins and Elmstrom (26) found no significant difference in wilt control of watermelons between high pH:all NO_3-N treatments and the low pH:NH_4-N treatments. However, they were farming a deep soil without a hardpan and the roots may have grown through the high pH top layer into a *Fusarium* -infested lower soil profile with a low pH. Pergola et al. (48) on carnation found that the severity of Fusarium wilt decreased when NH_4-N was used compared with NO_3-N. Nevertheless, a high soil pH of 7.5 resulted in less wilt.

POTASSIUM

The nitrogen (N) work of Walker and Foster (70) also shed light on the effect of potassium (K) on the development of Fusarium wilt of tomato. They found that when using all NO_3-N that low K favored disease development, whereas high K retarded it. In contrast, Foster and Walker (25) found that tomato plants preconditioned with low N or high K prior to inoculation were more susceptible than those preconditioned with high N or low K.

Schneider (54) reported inhibition of Fusarium yellows of celery with a specific concentration of potassium chloride (KCl) and NO_3-N. He further attributed this control to the ratio of K and Cl in the plant tissue. Ramasamy and Prasad (50) reported that increasing rates of KCl applied to muskmelon in pots led to a reduction in Fusarium wilt incidence. However, Spiegel and Netzer (60) and Stoddard (63) working with muskmelon, found that K level had very little effect on Fusarium wilt regardless of N source. Tharp and Wadleigh (66) demonstrated in a sand nutrient-culture experiment that an increase in K was accompanied by a significant reduction in severity of wilt in cotton. The severity of Fusarium wilt of cotton was decreased by increasing K rates in a field experiment by Dick and Tisdale (10).

PHOSPHORUS

Woltz and Jones (75-78) reported that a high level of phosphorus (P) increased the severity of Fusarium wilt of tomato in pot and field experiments and that the combination of high lime plus low P greatly curtailed disease development. They showed that supplemental applications of superphosphate (in field plots) above the amount required for growth of tomato greatly increased the occurrence of wilt in soils of pH 6.0. At pH 7.0 or 7.5 supplemental applications did not increase wilt occurrence because at high pH values P becomes sparingly available. Similarly, Sagdullaev and Berezhnova (51) reported that P amendments increased the severity of Fusarium wilt of muskmelon and Dick and Tisdale (10) found that increasing P rates increased the severity of Fusarium wilt of cotton.

TOTAL SALTS AND CHLORIDES

Walker and Foster (70) conducted nutritional experiments on Fusarium wilt of tomato using solution culture techniques and found that Fusarium wilt severity declined with an increase in salt concentration. Stoddard and Diamond (64), Mackay (43), and Standaert et al. (62) also reported development of tomato wilt to be inhibited with an increasing nutrient concentration.

Davet *et al.* (9) reported that a cool weather temperature strain of *F. oxysporum* f. sp. *lycopersici* existed in Morocco that produced severe wilt symptoms on tomato, especially if affected plants were irrigated with NaCl or MgCl$_2$. Schneider's (54) work on Fusarium yellows of celery also indicated that Cl salts favored disease development.

MICRONUTRIENTS

Fusarium oxysporum f. sp. *lycopersici* was shown by Jones and Woltz (33,36) and Woltz and Jones (73,77,78) to have a relatively high requirement for micronutrients. Deficiencies of Cu, Fe, Mn, molybdenum, and Zn reduced growth and sporulation of the fungus. Moreover, *F. oxysporum* f. sp. *lycopersici* grown in liquid cultures devoid of molybdenum or Zn was not as virulent as *F. oxysporum* f. sp. *lycopersici* grown on optimal levels of those two micronutrients. The response of *Fusarium* to Mn, Fe, and Zn was very pronounced. Increasing amounts of these micronutrients (above those usually found in soil solutions for tomato culture) were increasingly beneficial to growth and sporulation of the pathogen, whereas concentrations below average inhibited growth and spore production.

Liming soil infested with *F. oxysporum* f. sp. *oxysporum* in the field or greenhouse to pH 7.0-7.5 greatly limited the availability of micronutrients and consistently decreased wilt in naturally low pH soils. However, when soil with a high pH was further amended with lignosulfonate metal complexes of Zn, Mn, or Fe (the metal ions in these complexes are available for plant growth at high soil pH values), the beneficial effect of pH elevation was reversed according to Jones and Woltz (33). Consequently, the beneficial effects of liming apparently were not caused by an increased soil or tissue Ca content, but rather by the unavailability of micronutrients created by the high soil pH, which in turn limited the growth, sporulation, and virulence of the pathogen.

Iron in adequate supply was an important factor in the Fusarium wilt process according to Scher and Baker (53), Simeoni et al. (58), and Yuen and Schroth (82). It was suggested by Scher and Baker that the management of Fe availability through Fe competition can reduce wilt severity. Siderophores, produced by bacteria, complex Fe^{+++} iron and reduce its availability to *Fusarium*. Adding FeCl$_3$ to the soil overcame *Fusarium* suppression. A similar mechanism of Fe starvation for the Fusarium pathogen was shown by Jones and Woltz (32- 34) and Woltz and Jones (73) for the Fusarium wilt of tomato pathogens.

Edgington and Walker (14) in their work with Ca found that the influence of boron (B) on disease was dependent upon the Ca supply so that with 100 ppm Ca the disease decreased with increasing B, but at 500 ppm Ca, disease index increased with increases in B. Working with flax (*Linum usitatissimum* L.), Keane and Sackston (38) found that a B deficiency exacerbated development of Fusarium wilt (*F. oxysporum* Schlecht. f. sp. *lini* (Bolley) Snyder and Hansen).

Silicon was reported by Miake and Takahasi (44) to remarkedly lower the incidence of Fusarium wilt of cucumber.

POSSIBLE MODES OF ACTION OF LIME
AND NITRATE NITROGEN

High soil pH greatly limits the availability of micronutrients essential for the growth, sporulation, and virulence of the wilt fusaria. The work of Jones and Woltz (33) conclusively demonstrated that the inhibition of Fusarium wilt of tomato by high soil pH could be reversed by the addition of micronutrient compounds that were available at high soil pH values. Increases in soil pH limits not only the availability of micronutrients, but also the availability of other elements essential for *F. oxysporum*, especially phosphorus. Additionally, actinomycete and bacterial populations are favored by a high soil pH (68). Certain of these microorganisms are antagonostic to *F. oxysporum* f. sp. *lycopersici*, preventing spore germination and vegetative growth by means of toxic compounds. Bacteria and actinomycetes also compete with *Fusarium* for organic and inorganic nutrients in the soil solution.

The effect of N source may be fourfold: first, NO_3-N increases and NH_4-N decreases soil pH; second, *F. oxysporum* f. sp. *lycopersici* grown on NH_4-N is far more virulent than *Fusarium* grown on NO_3-N; third, tomato seedlings grown on NO_3-N are preconditioned and are more resistant to Fusarium wilt than seedlings grown on NH_4- N; and fourth, Schneider (54) attributed disease suppression by NO_3-N to a specific tissue concentration ratio of K:Cl. He reported disease was least when the K:Cl ratio was about 3.5, with more severe disease developing at ratios above or below 3.5. Factors affecting the ratio were N source, the availability of competing cations (Ca and NH_4) and anions (SO_4 and NO_3), and the concentration of available K and Cl in the soil. Ammonium repressed uptake of NO_3 and K and stimulated Cl uptake, thereby favoring disease development. The effect of N source on soil pH seems to be one of the most important factors.

CONCLUSIONS

Although the mechanism of control of Fusarium wilt with lime, NO_3-N, and low P is not known with certainty, there is little doubt that the regime is highly effective. Despite a few conflicting observations in the literature, certain consistencies are apparent. First, liming the soil to a pH of 6.0 to 7.5, in the absence of organic matter or NH_4-N, routinely decreased the incidence and severity of Fusarium wilt of many crops. In fact, Jones and Overman (29) found that raising the pH of EauGallie fine sand to 7.0 controlled tomato wilt as well as pasteurizing pH 5.5 soil with a broad-spectrum soil fumigant. Second, N source almost invariably affects disease development. The level of wilt decreases when NO_3-N is used in the fertilizer and increases when NH_4-N is used. Finally, high P encourages wilt development.

Engelhard and Woltz (16-18), by combining the lime:NO_3-N system with benomyl, obtained complete control of Fusarium wilt of chrysanthemum and aster. The lime:NO_3-N and the lime:NO_3-N:chemotherapy systems have proven to be eminently successful in obtaining a high level of control of Fusarium wilt of tomato, cucumber, watermelon, gladiolus, chrysanthemum, aster, carnation, and other crops not only in experimental plots in Florida but throughout the world in commercial fields and plantings, as well as in experimental plots. For example, the system was found effective in commercial chrysanthemum ranges in Florida (49) and was reported to be effective in Maryland (40). Sarhan and Kiraly (52) reported complete control of Fusarium wilt of potted tomatoes in Hungary by using a high soil pH:NO_3-N:benomyl system. Jones and Woltz (37) found the lime:NO_3-N regime gave excellent control of Fusarium wilt of tomato in Florida caused by a new pathogenic race. The combination of NO_3-N and benomyl resulted in excellent control of Fusarium wilt of Hebe in California. The use of $Ca(NO_3)_2$ and a high soil pH also resulted in control of Fusarium wilt of carnation in England (11). Undoubtedly, the systems consistently are effective in the control of Fusarium wilt of many crops throughout the world.

LITERATURE CITED

1. Ahmet, K. 1933. Untersuchen uber tracheomykosen.Phytopath. Zeitschr. 6:49-101.
2. Albert, W.B. 1946. The effects of certain nutrient treatments upon the resistance of cotton to *Fusarium vasinfectum*. Phytopathology 36:703-716.
3. Armstrong, G.M., and Armstrong, J.K. 1964. Wilt of

chrysanthemum caused by race 1 of the cowpea Fusarium. Phytopathology 54:886 (Abstr.).

4. Armstrong, G.M., Armstrong, J.K., and Littrell, R.H. 1970. Wilt of chrysanthemum caused by *Fusarium oxysporum* f. sp. *chrysanthemi*, forma specialis nov. Phytopathology 60:496-498.

5. Bloom, J.R., and Walter, J.C. 1955. Effect of nutrient sprays on Fusarium wilt of tomato. Phytopathology 45:443-444.

6. Clayton, E.E. 1923. The relation of soil moisture to the Fusarium wilt of the tomato. Am. J. Bot. 10:133-147.

7. Cook, W.S. 1937. Relation of nutrition of tomato to disposition to infectivity and virulence of *Fusaruim lycopersici*. Bot. Gaz. 98:647-669.

8. Corden, C.E. 1965. Influence of calcium nutrition on Fusarium wilt of tomato and polygalacturonase activity. Phytopathology 55:222-224.

9. Davet, P., Messiaen, C.M., and Rieuf, P. 1966. Interpretation of winter manifestation of Fusarium wilt of tomato in North Africa, favored by irrigation water salts. (Trans. title in French). Proc. First Cong. Mediterr. Phytopath. Union:407- 416.

10. Dick, J.B., and Tisdale, H.B. 1938. Fertilizers in relation to incidence of wilt as affecting a resistant and a susceptible variety. Phytopathology 28:666-667 (Abstr.).

11. Ebben, M.H. 1979. Carnation wilt caused by *Fusarium oxysporum* f. sp. *dianthi*. Ann. Rep. Glasshouse Crops Res. Inst. 207 pp.

12. Edgerton, C.W., and Moreland, C.C. 1913. Diseases of the tomato in Louisiana. La. Agric. Exp. Stn. Bull. 142.

13. Edgerton, C.W., and Moreland, C.C. 1920. Tomato wilt. La. Agric. Exp. Stn. Bull. 174.

14. Edgington, L.V., and Walker, J.C. 1958. Influence of calcium and boron nutrition on development of Fusarium wilt of tomato. Phytopathology 48:324-326.

15. Engelhard, A.W. 1975. Aster Fusarium wilt: Complete symptom control with an integrated fungicide-NO_3-pH control system. Proc. Am. Phytopathol. Soc. 2:62 (Abstr.).

16. Engelhard, A.W. 1979. Control of Fusarium wilt of carnation with an integrated (NO^-_3)-N and systemic fungicide control program. Phytopathology 69:1027 (Abstr.).

17. Engelhard, A.W., and Woltz, S.S. 1972. Complete control of Fusarium wilt of chrysanthemum with chemotherapeutants combined with a high lime and nitrate-nitrogen culture regime. Phytopathology 62:756.

18. Engelhard, A.W., and Woltz, S.S. 1973. Fusarium wilt of chrysanthemum: Complete control of symptoms with an

integrated fungicide-lime-nitrogen regime. Phytopathology 63:1256-1259.

19. Engelhard, A.W., and Woltz, S.S. 1978. Effect of temperature, nitrogen source, lime and benomyl treatments on Fusarium wilt of chrysanthemum, aster, and gladiolus. Third Int. Congr. of Plant Pathology p. 375 (Abstr.).

20. Essary, S.H. 1912. Notes on tomato diseases with results of selection for resistance. Tenn. Agric. Exp. Stn. Bull. 95.

21. Essary, S.H. 1920. Report of the botanist. Tenn. Agric. Exp. Stn. Annu. Rep. 1919-20:15-16.

22. Everett, P.H., and Blasquez, C.H. 1967. Influence of lime on the development of Fusarium wilt of watermelon. Proc. Fla. State Hort. Soc. 80:143- 148.

23. Fisher, P.L. 1935. Physiological studies on the pathogenicity of *Fusarium lycopersici* Sacc. for the tomato plant. Md. Agric. Exp. Stn. Bull. 374.

24. Fisher, P.L. 1935. Responses of tomato in solution cultures with deficiencies and excesses of certain essential elements. Md. Agric. Exp. Stn. Bull. 375.

25. Foster, R.E., and Walker, J.C. 1947. Predisposition of tomato to Fusarium wilt. J. Agric. Res. 74:165-185.

26. Hopkin, D.L., and Elmstrom, G.W. 1976. Effect of soil pH and nitrogen source on Fusarium wilt of watermelon on land previously cropped to watermelons. Proc. Fla. State Hort. Soc. 89:141- 143.

27. Huang, J.W., and Sun, S.K. 1982. The effects of nitrogenous fertilizers on disease development of watermelon fusarial wilt. Plant Prot. Bull. (Taiwan ROC) 24:101-110.

28. Huang, J.W., Sun, S.K., and Juang, C.F. 1986. Studies on the integrated control of radish yellows, caused by *Fusarium oxysporum* f. sp.*raphani*. Plant Prot. Bull. (Taiwan ROC) 28:81-90.

29. Jones, J.P., and Overman, A.J. 1971. Control of Fusarium wilt of tomato with lime and soil fumigants. Phytopathology 61:1415-1417.

30. Jones, J.P., and Woltz, S.S. 1967. Fusarium wilt (race 2) of tomato: effect of lime and micronutrient soil amendments on disease development. Plant Dis. Rep. 51:645-648.

31. Jones, J.P., and Woltz, S.S. 1968. Field control of Fusarium wilt (race 2) of tomato by liming and stake disinfestation. Proc. Fla. State Hort. Soc. 81:187-191.

32. Jones, J.P., and Woltz, S.S. 1969. Fusarium wilt (race 2) of tomato: calcium, pH, and micronutrient effects on disease development. Plant Dis. Rep. 53:276-279.

33. Jones, J.P., and Woltz, S.S. 1970. Fusarium wilt of tomato: Interaction of soil liming and micronutrients on disease development. Phytopathology 60:812-813.
34. Jones, J.P., and Woltz, S.S. 1972. Effect of soil pH and micronutrient amendments on Verticillium and Fusarium wilt of tomato. Plant Dis. Rep. 56:151-153.
35. Jones, J.P., and Woltz, S.S. 1975. Effect of liming and nitrogen source on Fusarium wilt of cucumber and watermelon. Proc. Fla. State Hort. Soc. 88:200-203.
36. Jones, J.P., and Woltz, S.S. 1981. Fusarium-incited diseases of tomato and potato and their control. Pages 157-168 in: Fusarium: Diseases, Biology and Taxonomy. P. E. Nelson, T. A. Toussoun and R. J. Cook, eds. Penn. State Univ. Press, University Park.
37. Jones, J.P., and Woltz, S.S. 1983. Cultural control of Fusarium wilt race 3 of tomato. Proc. Fla. State Hort. Soc. 96:82-83.
38. Keane, E.M., and Sackston, W.E. 1970. Effects of boron and calcium nutrition of flax on Fusarium wilt. Can. J. Plant Sci. 50:415-422.
39. Keim, R., and Humphrey, W.A. 1984. Fertilizer helps control Fusarium wilt of Hebe. Calif. Agric. 38:12-14.
40. Locke, J.C., Marois, J.J., and Papavizas, G.C. 1985. Biological control of Fusarium wilt of greenhouse-grown chrysanthemums. Plant Dis. 69:167-169.
41. Loffler, H.J.M., Cohen, E.B., Oolbekkink, G.T., and Schippers, B. 1986. Nitrite as a factor in the decline of *Fusarium osysporum* f. sp. *dianthi* in soil supplemented with urea or ammonium chloride. Neth. J. Plant Path. 92:153-162.
42. Loffler, H.J.M., Hoelman, A., Nielander, H.B., and Schippers, B. 1986. Reduced chlamydospore formation and enhanced lysis of chlamydospores of *Fusarium oxysporum* in soil with added urea or ammonium chloride. Biol. Fert. Soils 2:1-6.
43. Mackay, J.H.E. 1952. Fusarium wilt of tomato-the effects of level of nutrition on disease development. J. Aust. Inst. Agric. Sci. 17:207- 211.
44. Miyake, Y., and Takahashi, E. 1983. Effect of silicon on the growth of cucumber plants in soil culture. Soil Sci. Plant Nutr. 29:463-471.
45. Morgan, K.T., and Timmer, L.W. 1984. Effect of inoculum density, nitrogen source and saprophytic fungi on Fusarium wilt of Mexican lime. Plant Soil 79:203-210.
46. Norton, J.B.S. 1912. Differences in varieties of fruit and truck crops in reference to disease. Rep. Md. State Hort. Soc. 15:62-67.
47. Norton, J.B.S. 1914. Tomato diseases. Md. Agric. Exp. Stn. Bull.

180:102-114.
48. Pergola, G., Guda, C.D., and Garibaldi, A. 1979. Fusarium wilt of carnation: Effect of soil pH and N source on disease development. Med. Fac. Landbouww Rijksuniv Gent. 44(1):414-421.
49. Raju, B.C. 1983. Fusarium wilt of mums. Acta Hort. 152:65-76.
50. Ramasamy, K., and Prasad, N.N. 1975. Effect of potassium nutrition on phenol metabolism of melon wilt. Madras Agric. J. 62:313-317.
51. Sagdullaev, M.M., and Berezhnova, W. 1974. The effects of phosphorus fertilizers on the physiological properties, yield, and resistance of melons to fusarium wilt. AgroKhimiya 4:36-40.
52. Sarhan, A.R.T., and Kiraly, Z. 1981. Control of Fusarium wilt of tomato with an integrated nitrate-lime-fungicide regime. Acta Phytopathol. Acad. Sci. Hung. 16:9-14.
53. Scher, F.M., and Baker, R. 1982. Effect of *Pseudomonas putida* and a synthetic iron chelator on induction of soil suppressiveness to Fusarium wilt pathogen. Phytopathology 72:1567-1573.
54. Schneider, R.W. 1985. Suppression of Fusarium yellows of celery with potassium, chloride, and nitrate. Phytopathology 75:40-48.
55. Scott, I.T. 1924. The influence of hydrogen ion concentration on the growth of *Fusarium lycopersici* and on tomato wilt. Mo. Agric. Exp. Stn. Res. Bull. 64. 32 pp.
56. Sherbakoff, C.D. 1915. Fusaria of potatoes. Cornell Univ. Agric. Exp. Stn. Memoir 6.
57. Sherwood, E.C. 1923. Hydrogen-ion concentration as related to the Fusarium wilt of tomato seedlings. Am. J. Bot. 10:537-553.
58. Simeoni, L.A., Lindsay, W.L., and Baler, R. 1987. Critical iron level associated with biological control of Fusarium wilt. Phytopathology 77:1057- 1061.
59. Smith, E.F. 1899. Wilt diseases of cotton, watermelon, and cowpea. US Dept. Agric. Bur. Plant. Ind. Bull. 17.
60. Spiegel, Y., and Netzer, D. 1984. Effect of nitrogen form at various levels of potassium on the Meloidogyne-Fusarium wilt complex in muskmelon. Plant and Soil 81:85-92.
61. Standaert, J.Y., Myttenaere, C., and Meyer, J.A. 1973. Influence of sodium/calcium ratios and ionic strength of the nutrient solution on Fusarium wilt of tomato. Plant Sci. Lett. 1:413- 420.
62. Standaert, J., Maraite, H., Myttenaere, C., and Myer, J.A. 1978. A study of the effect of salt concentration and sodium/calcium ratio in the nutrient medium on the susceptibility of tomatoes to fusarium wilt. Plant Soil 50:269-286.
63. Stoddard, D.L. 1947. Nitrogen, potassium,and calcium in relation to Fusarium wilt of muskmelon. Phytopathology 37:875-

884.

64. Stoddard, E.M., and Diamond, A.E. 1948. Influence of nutritional level on the susceptibility of tomatoes to Fusarium wilt. Phytopathology 38:670- 671.

65. Sun, S.K., and Huang, J.W. 1985. Formulated soil amendment for controlling Fusarium wilt and other soilborne diseases. Plant Dis 69:917-920.

66. Tharp, W.H., and Wadleigh, C.H. 1939. The effects of nitrogen phosphorus, and potassium nutrition on the Fusarium wilt of cotton. Phytopathology 29:756.

67. Trillas-Gay, M.I., Hoitink, H.A.J., and Madden, L.V. 1986. Nature of suppression of Fusarium wilt of radish in a container medium amended with composted hardwood bark. Plant Dis. 70:1023-1027.

68. Waksman, S.A. 1927. Principles of soil microbiology. The Williams and Wilkins Co., Baltimore, MD. 897 pp.

69. Walker, J.C. 1971. Fusarium wilt of tomato. The Am. Phytopathol. Soc. Monograph No. 6. Am. Phytopathol. Soc., St. Paul, MN.

70. Walker, J.C., and Foster, R.E. 1946. Plant nutrition in relation to disease development. III Fusarium wilt of tomato. Am. J. Bot. 33:259-264.

71. Wollenweber, H.W. 1913. Studies on the Fusarium problem. Phytopathology 3:24-50.

72. Woltz, S.S., and Engelhard, A.W. 1973. Fusarium wilt of chrysanthemum: Effect of nitrogen source and lime on disease development. Phytopathology 63:155-157.

73. Woltz, S.S., and Jones, J.P. 1968. Micronutrient effects on the *in vitro* growth and pathogenicity of *Fusarium oxysporum* f. sp. *lycopersici*. Phytopathology 58:336-338.

74. Woltz, S.S., and Jones, J.P. 1972. Control of Fusarium wilt of tomato by varying the nutrient regimes in the soil. Phytopathology 62:799. (Abstr.).

75. Woltz, S.S., and Jones, J.P. 1973. Interactions in source of nitrogen fertilizer and liming procedure in the control of Fusarium wilt of tomato. Hort Science 8:137-138.

76. Woltz, S.S., and Jones, J.P. 1973. Tomato Fusarium wilt control by adjustments in soil fertility. Proc. Fla. State Hort. Soc. 86:157-159.

77. Woltz, S.S., and Jones, J.P. 1973. Tomato Fusarium wilt control by adjustments in soil fertility: A systematic approach to pathogen starvation. Agric. Res. Ed. Center Bradenton Res. Rept. GC1973-7.

78. Woltz, S.S., and Jones, J.P. 1981. Nutritional requirements of

Fusarium oxysporum: Basis for a disease control system. Pages 340-349 in: Fusarium: Diseases, Biology, and Taxonomy. P.E. Nelson, T. A. Toussoun, and R.J. Cook, eds. Penn. State Univ. Press, University Park.

79. Woltz, S.S., and Jones, J.P. 1984. Effects of aluminum, lime, and phosphate combinations on Fusarium wilt (race 3) of tomato. Phytopathology 74:629 (Abstr.).

80. Woltz, S.S., and Magie, R.O. 1975. Gladiolus Fusarium disease reduction by soil fertility adjustments. Proc. Fla. State Hort. Soc. 88:559- 562.

81. Yoshino, M., and Hashimoto, K. 1978. Studies on the ecology of strawberry yellows and its control. Bull. Saitama Hort. Exp. Stn. No 7. p. 13-34.

82. Yuen, G.Y., and Schroth, M.N. 1986. Inhibition of *Fusarium oxysporum* f. sp. *dianthi* by Fe competition with an *Alcaligenes* sp. Phytopathology 76:171-176.

THE ROLE OF MINERAL NUTRITION IN THE CONTROL OF VERTICILLIUM WILT

Barbara W. Pennypacker
Department of Plant Pathology
The Pennsylvania State University
University Park, Pennsylvania 16802

Vascular wilt diseases are among the most destructive plant diseases and have the potential to severely limit yield in many economically important crops. Members of the genera *Fusarium* and *Verticillium* are responsible for a major portion of these diseases, with *Verticillium* being a problem mainly in the cooler, temperate regions of the world. Of the five species of *Verticillium* known to be vascular pathogens, *V. albo-atrum* and *V. dahliae* are the most important (17). *Verticillium albo-atrum* Reinke. et Berth. and *V. dahliae* Kleb. attack a wide range of ornamentals, vegetables, fruits, forages and forest trees (30,36,42). These two soilborne pathogens differ primarily in the resting structures they form. *Verticillium dahliae* produces microsclerotia, which can survive in the soil up to 8 years (42), while *V. albo-atrum* forms melanized resting hyphae, which remain viable 11 to 13 months in plant debris (29,32). The two pathogens have very limited saprophytic ability in the soil (31,32,42,51) but are capable of infecting a large number of weed species (20,21,42,48), thereby surviving periods of crop rotation to non-host species.

Verticillium wilt is considered to be a single cycle disease because the pathogen usually fails to produce inoculum that is effective in the same year (42). Measures aimed at reducing the amount of initial inoculum should, therefore, be effective in controlling the disease. Unfortunately, this is not true in the case of Verticillium wilt of alfalfa. *Verticillium albo-atrum* sporulates profusely on dead alfalfa stubble. Isaac and Heale (29) found that cutting hay effectively spread the pathogen through the crop. Alfalfa is harvested three to eleven times a growing season, depending on geographic location, thus, the alfalfa strain of *V. albo-atrum* has, in reality, a corresponding number of repeating cycles. Control of Verticillium wilt also is difficult in other perennial crops, because of yearly inoculum increases.

Varying degrees of control of Verticillium wilt have been achieved through sanitation, crop rotation, soil fumigation and soil solarization (10,25,30,32,42,46,50). The longevity of *V. dahliae* microsclerotia, the wide host range and the ability to infect weed species severely diminishes the effectiveness of crop rotation in controlling this pathogen (25). Crop rotation has had some success against *V. albo-atrum* in England, where infested hop gardens are rotated to grass for a minimum of 2 years before being replanted with hops (46). Undoubtedly, the limited persistence of the resting hyphae of *V. albo-atrum* is a major factor in the reduction of inoculum by this method. As is the case with *V. dahliae*, the wide host range of *V. albo-atrum* makes crop rotation with other dicotyledons difficult (48).

Due to the expense of soil fumigation and the limited success of crop rotation, resistance has been the traditional control method for Verticillium wilt diseases (36,42,51). In almost all host species, however, true immunity to *Verticillium* is lacking (49). Although dominant resistance has been identified in tomato, cotton, alfalfa, mint, sunflower and strawberry, in these and other cases, polygenic resistance is also operative (6,16,51). Resistance mediated by polygenes is known to be modified by environmental factors such as temperature, soil moisture, light and soil fertility (6,37), thus cultural manipulation can support the existing genetic resistance to *Verticillium*.

MINERAL NUTRITION AND VERTICILLIUM WILT

Soil fertility can affect Verticillium wilt diseases in two major ways, 1) by reducing inoculum density and, 2) by altering the expression of genetic resistance (6,23,24). In either case it is important to determine the initial level of fertility, because, for many minerals, altering the nutritional level of the soil only has an effect on resistance when the nutrient to be added is in the deficiency range or marginally adequate (35). In addition, the inoculum density should be assessed, because fertility manipulation cannot alter resistance when the host plants are exposed to extremely high levels of inoculum (3,42).

Effect of Nutrition on Inoculum Density

Propagules of *Verticillium* require stimulation from host exudates for germination (19,35,42) and the composition of these exudates is

34

governed, in part, by host nutrition. Soil amendments also may have a direct effect on the viability of the pathogen resting structures.

Wilhelm (47) found that adding bone or fish meal or $(NH_4)_2SO_4$ to naturally infested soil consistently reduced the inoculum potential of *V. albo-atrum* as measured by infection of susceptible tomato plants. It is difficult, however, to determine whether the observed effect was due to an effect on the pathogen or on the tomato indicator plants, because NH_4-N has since been found to increase resistance to *Verticillium* in several hosts (14,27,42). Duncan and Himelick (13) and Dutta and Isaac (14) observed the response of *Verticillium dahliae* to NH_4-N in vitro and found that hyphal growth was severely distorted on media amended with NH_4-N as compared to growth with other N sources. This finding supports Wilhelm's conclusions concerning the negative effect of this nutrient on inoculum potential.

Jordan et al. (31) noted reduced germination of *V. dahliae* microsclerotia and reduced mycelial growth when the soil was treated with chitin, laminarin, wheat straw or oven dried green clover. Addition to the soil of materials with a high C:N ratio results in the immobilization of N during microbial decomposition (4), and this reduction in N level may be responsible for the reduced germination observed. Reduced germination also can be due to the sorption of microsclerotia and metal ions in the soil solution. Ashworth et al. (1) attributed a sharp reduction in microsclerotia in cotton fields in California, as detected by plating methods, to fungistasis caused by Cu sorbed to the microsclerotia. When Cu was removed by treating the microsclerotia with either a metal chelating agent, or NaOCl or with monovalent desorbing agents, germination occurred. The authors also tested Fe, Mn, and Zn for fungistatic activity but were unable to demonstrate any effect on microsclerotial germination at ion concentrations compatible with those found in the San Joaquin Valley soils. The importance of metal ions as fungistatic agents thus appears to depend on their concentration and this will vary with soil type and pH (4).

Nutrition and Resistance to Verticillium Wilt

Much of the research into the effect of soil fertility on resistance to Verticillium wilt has been concentrated on the major elements, N, K and P. Conflicting reports abound in the literature concerning the effects of these elements on resistance (Table 1). In many cases, the initial level of soil fertility, the pH of the soil, the cation exchange capacity of the soil, and soil moisture status are not given. The pH of a soil determines the solubility of several ions including P, Ca, Fe and Al

Table 1. Effect of N, P, and K nutrition on Verticillium wilt.

Element	Level	Increased Resistance	Reduced Resistance
N	High	Antirrhinum (14,26) Cotton (6) Potato (9,10,22,23) Tomato (47)	Alfalfa (28) Antirrhinum (27) Cotton (37) Eggplant (45) Hop (34) Tomato (30,40)
N	Low	Alfalfa (28) Eggplant (45) Hop (34,44,46)	Potato (10)
P	High		Cotton (12)
P	Low	Cotton (12) Hops (34)	
K	High	Cotton (12,42) Hops (34) Pistachio (2,3)	Alfalfa (28)
K	Low		Cotton (12,18) Pistachio (2,3) Tomato (40)

and the number of pH-dependent exchange sites contributing to the cation exchange capacity (C.E.C.) of the soil. Cation exchange capacity, which involves the bonding of cations to soil particles and organic matter, influences the availability of Mg, Ca, Fe, B, Mn, K, NH_4-N and Zn. The C.E.C. varies depending on the type and amount of clay mineral and organic matter present in the soil (4). Soil pH, C.E.C., and the amount of moisture present in the soil, a factor affecting diffusion as well as mass flow of ions in the soil, influence the availability of nutrients added to the soil and, subsequently, the quantity of those nutrients taken up by the plant (4). Many of the conflicting reports concerning the effect of various nutrients on resistance to Verticillium wilt are probably due to difference in these soil parameters and to differences in inoculum density.

These problems prevent direct comparison among published results of studies on the effect of nutrition on resistance to Verticillium wilt. Results could, however, be compared if tissue analyses for the nutrients under study were reported. Unfortunately, few studies include such information. In most of the cases where tissue analysis was reported (3,9,10,12,15,18,34), good correlations were obtained between nutrient status and wilt resistance. Thus, by allowing the plant to integrate the soil factors affecting nutrient availability and the nutrient requirements for altering host resistance, we can compare the results from studies conducted under widely different conditions. Of course, host tissue analysis for mineral levels will not correlate with increased resistance that is due to a reduction in inoculum potential as previously mentioned.

Nitrogen

The summary of the literature concerning the effect of N on resistance to Verticillium wilt is shown in Table 1. Most of the reports included in this summary were based on correlations between visual disease assessment and the amount of nutrient added to the soil. Few include analyses for the concentration of these nutrients in the host tissue. Nitrogen has the greatest effect on this host-pathogen system. It also is clear that there is no consensus concerning the effect of this element on resistance. Although some of the conflicting results may be attributed to the reasons previously mentioned, a comparison of results when inorganic N source was specified indicates that differences in the source of N are also involved (23,24).

Table 2 includes, with one exception, results from studies where either NH_4-N or NO_3-N fertilizers were applied to the soil, and it is clear that the effect on resistance to *Verticillium* differed accordingly. Davis and Everson (10), working with Verticillium wilt of potato, noted a positive correlation between NO_3-N level in the soil at season's end and incidence of wilt (r=0.629, p=.001), and a negative correlation between petiole NO_3-N levels and wilt (r=-0.544, p=.01). This apparent conflict in the effect of NO_3-N on wilt incidence may reflect the use of NH_4-N nitrate fertilizer in the study, thus, both NH_4-N and NO_3-N were supplied to the host. It is generally believed that reduced N rates favor resistance, however, NH_4-N was found to increase resistance in Antirrhinum, potato, and possibly tomato (14,23,27,47) and to delay symptom expression in alfalfa (28).

Table 2. Effect of soil N source on resistance to Verticillium wilt.

Nitrogen	Level	Increased Resistance	Reduced Resistance
NH_4-N	High	Alfalfa (28) Antirrhinum (14,26) Potato (23) Tomato (46)	Hop (44)
NH_4-N	Low	Hop (44)	
NO_3-N	High		Antirrhinum (27) Potato (10) Tomato (30,40)
NO_3-N	Low	Cotton (39) Hop (34)	

Differences in response of the *Verticillium* infected plant to the two N sources are not unexpected because NH_4-N and NO_3-N have different effects on the cation-anion balance in the plant. NH_4-N inhibits cation uptake and can, consequently, cause deficiencies in Ca and Mg uptake. In addition, the rhizosphere pH is altered when NH_4-N is taken up due to the necessary extrusion of H^+ to balance the charge in the cytoplasm. NO_3-N, on the other hand, causes an influx of cations and a resulting increase in rhizosphere pH when OH^- or HCO_3^- are extruded to balance the charge (4,35). These differences in ion uptake by the plant alter the ion balance in the plant and therefore would affect the expression of resistance that depends on host chemical composition and metabolic activity.

The two N sources also cause differences in the levels of soluble carbohydrates in the root. Due to its inherent toxicity to the plant, NH_4-N is immediately incorporated into amino acids and amides when it enters a root cell. This process requires a large supply of carbohydrates to provide the necessary carbon skeletons and consequently increases respiration and reduces the amount of soluble carbohydrates. Nitrate, on the other hand, may be stored in the cell vacuole or may be reduced to NH_3 in the root cell. Nitrate reduction also requires carbon skeletons and therefore reduces the level of soluble carbohydrates in the roots. However, when the level of NO_3-N in the soil is high, NO_3-N is translocated in the xylem to the shoots where it is then reduced. Thus, under high levels of NO_3-N fertilization, levels of root soluble carbohydrate will be higher than they are when NO_3-N levels are low (35).

Roberts (41) conducted an interesting experiment during which the fertility level of the soil was kept at adequate levels and the carbohydrate level of the plant was reduced by altering the leaf:shoot ratio via defoliation. Tomato plants acquired increased resistance to *Verticillium albo-atrum* when their carbohydrate levels were reduced. It may be possible that the varying effects of the NO_3-N and NH_4-N sources on resistance to *Verticillium* also could be attributed to differing levels of root carbohydrate.

Potassium

The effect of K on resistance to *Verticillium* is only evident when the element is deficient in the soil (35). This may explain the limited number of reports of increasing resistance by increasing K fertility levels (Table 1).

Ashworth (2,3) found that pistachio trees, which were deficient in K, had a much higher incidence of infection by *V. dahliae* than did trees with adequate K. When the trees were growing in soil with an inoculum density of only 0.02 to 0.2 microsclerotia per gm soil, 39.6% of the K-deficient trees were infected vs. 0.37% of the K-sufficient trees. Applications of K at the rate of 1.5 kg of K per tree resulted in a 35% reduction in the number of infected trees. Ashworth et al. (3) felt the increased level of resistance was due to improved root growth which shortened the time the susceptible root tips were exposed to the microsclerotia. The authors noted, however, that improved nutrition was of no practical significance when the inoculum density was high ($\geq$ 5 microsclerotia/gm soil).

Hafez et al. (18), working with Verticillium wilt of cotton, found a similar relationship between K deficiency and resistance to *V. dahliae*. Again, increased levels of K increased the level of resistance to the vascular pathogen. The opposite situation, however, was found in alfalfa infected with *V. albo-atrum*. In this case, Isaac (28) noted that increased levels of K_2SO_4 actually reduced time until symptom expression by 50%. The incidence of wilt was not affected by the addition of K.

Potassium is not a structural component of plants but is rather a mobile charge carrier. It is involved in enzyme activation, membrane transport mechanisms, protein synthesis and cell extension. It is clear that imbalances in this element have the potential to alter the host metabolism and impair its ability to respond to pathogen invasion (22,35). Hafez et al. (18) noted that the addition of potassium to the deficient soils not only reduced the incidence of wilt but also decreased its severity. It is possible that improved root elongation was responsible for the reduction in the incidence of Verticillium wilt and an improved ability to respond to pathogen invasion was responsible for the reduction in severity.

Phosphorus

Alteration of P fertility levels has met with little success in increasing levels of resistance to *Verticillium*. Davis et al. (12) found that adequate levels of P in cotton plants, whether achieved by increased levels of P in the soil or by improved P uptake due to mycorrhizal associations, resulted in an increase in wilt due to *V. dahliae*. Phosphorus is a relatively immobile element, thus, mycorrhizae improve plant uptake of this element by increasing the soil area from which the root can extract the ion. Davis et al. (11) observed a decrease in *V. dahliae* colonization of potato stems when both N and P levels were optimal but no significant differences when the two nutrients were optimized separately. Isaac (28) noted increased incidence of wilt in alfalfa grown under high levels of superphosphate. The situation in alfalfa was unique, however, because the increase in disease incidence only occurred when the pathogen was *V. dahliae*, a weakly virulent pathogen of alfalfa. No response was noted when the alfalfa pathogen was the highly virulent *V. albo-atrum*. Similarly, no response was seen in either tomato (40) or cacao (15) when P levels were altered.

Calcium

The effect of Ca on resistance to Verticillium wilt also is unclear, due in part to the confounding effect of liming and the subsequent alteration in pH on the availability of other nutrients. Selvarja (43), working with eggplant and tomato, noted increased Verticillium wilt resistance when Ca levels were enhanced. He subsequently studied the effect of Ca nutrition on the growth of *V. dahliae* by supplementing the growing medium with $CaCl_2$. Calcium chloride, up to a concentration of 0.05 M, increased the production of endopolygalacturonase, unlike the situation in Fusarium wilt of tomato (8), where the enzyme was inhibited by increased Ca levels. Cooper et al. (7) found that the addition of 0.01 M Ca to a reaction mixture containing cell wall material and endopolygalacturonase reduced the optimal level of enzyme activity by 20%. In all of these experiments, the levels of Ca far exceeded the levels of this ion normally present in plants (35). It appears, then, that the effect of Ca is not directly on the production or activity of this pectic enzyme.

Pathogen produced endopolygalacturonase degrades pectic components of the xylem vessel wall, and the resulting pectic fragments react with divalent cations, especially Ca, to form gels which then block the vessel (38). Timely xylem vessel occlusion has been shown to limit the spread of vascular wilt pathogens through the plant (38), and such a defense mechanism, rather than a direct effect on enzyme activity, may be responsible for the enhanced resistance Selvaraj (43) noted. It also is possible that increased resistance to Verticillium wilt with increased levels of Ca (43) could be due to the conversion of pectin to Ca-pectate in the host. Marschner (35) noted that plants grown under high levels of Ca showed a disproportionate amount of Ca-pectate in their cell walls. Bateman and Lumsden (5), working with *Rhizoctonia solani* in beans, found that resistance was associated with increased levels of Ca-pectate, a form of pectin that is not hydrolyzed by endopolygalacturonase. Thus, regardless of the effect of Ca on the activity or production of this enzyme, conversion of the pectic materials in the cell wall to Ca-pectate would increase their resistance to enzymatic degradation.

CONCLUSION

Altering the nutritional state of the soil can have an effect on the severity of Verticillium wilt, particularly when the nutrient being added is in the deficient range of availability. In general, levels of N, which are adequate for good plant growth but do not promote luxury consumption, enhance resistance. Nitrate-N appears to increase susceptibility to *Verticillium*, but the effect of NH_4-N on host response varies with host species. Since the effect of fertility on resistance to *Verticillium* is governed by the interaction of the host, the pathogen and the environment, it is infinitely variable, and thus precludes specific recommendations.

ACKNOWLEDGEMENTS

Appreciation is expressed to the Department of Plant Pathology, The Pennsylvania State University and to the U.S. Regional Pasture Research Laboratory, USDA-ARS, for support received during the preparation of this manuscript.

LITERATURE CITED

1. Ashworth, L.J., Jr., Huisman, O.C., Grogan, R.G., and Harper, D.M. 1976. Copper-induced fungistasis of microsclerotia of *Verticillium albo-atrum* and its influence on infection of cotton in the field. Phytopathology 66:970-977.

2. Ashworth, L.J., Gaona, S.A., and Surber, E. 1985. Verticillium wilt of pistachio: the influence of potassium nutrition on susceptibility to infection by *Verticillium dahliae*. Phytopathology 75:1091-1093.

3. Ashworth, L.J., Morgan, D.P., and Surber, E. 1986. Verticillium wilt of pistachio. Calif. Agric. July-Aug. 1986:21-24.

4. Barber, S.A. 1984. Soil nutrient bioavailability. John Wiley & Sons, NY. 398 p.

5. Bateman, D.F., and Lumsden, R.D. 1965. Relation of calcium content and nature of the pectic substances in bean hypocotyls of different ages to susceptibility to an isolate of *Rhizoctonia solani*. Phytopathology 55:734-738.

6. Bell, A.A. 1973. Nature of disease resistance. National Cotton Pathology Research Lab. College Station, TX. ARS-S19. p. 47-62.

7. Cooper, R.M., Rankin, B., and Wood, R.K.S. 1978. Cell wall-degrading enzymes of vascular wilt fungi. II. Properties and mode of action of polysaccharidases of *Verticillium albo-atrum* and *Fusarium oxysporum* f. sp. *lycopersici*. Physiol. Plant Pathol. 13:101-134.

8. Corden, M.E. 1965. Influence of calcium nutrition on Fusarium wilt of tomato and polygalacturonase activity. Phytopathology 55:222-224.

9. Davis, J.R. 1985. Approaches to control of potato early dying caused by *Verticillium dahliae*. Am. Potato J. 62:177-185.

10. Davis, J.R., and Everson, D.O. 1986. Relation of *Verticillium dahliae* to soil and potato tissue, irrigation method, and N-fertility to Verticillium wilt of potato. Phytopathology 76:730-736.

11. Davis, J.R., Stark, J.C., and Sorensen, L.H. 1988. Reduced *Verticillium dahliae* colonizations in potato stems, wilt suppression, and reductions of *V. dahliae* cfu in soil with optimal N and P, 1986. Biological & Cultural Tests, American Phytopathological Society, Vol. III (in press).

12. Davis, R.M., Menge, J.A., and Erwin, D.C. 1979. Influence of *Glomus fasciculatus* and soil phosphorous on Verticillium wilt of cotton. Phytopathology 69:453-456.

13. Duncan, D.R., and Himelick, E.B. 1986. Inhibition of conidial production of *Verticillium dahliae* with ammonium sulfate. Phytopathology 76:788-792.

14. Dutta, B.K., and Isaac, I. 1979. Effects of inorganic amendments (N, P and K) to soil on the rhizosphere microflora of Antirrhinum plants infected with *Verticillium dahliae* Kleb. Plant and Soil 52:561-569.

15. Emechebe, A.M. 1980. The effect of soil moisture and of N, P and K on incidence of infection of cacao seedlings inoculated with *Verticillium dahliae*. Plant and Soil 54:143-147.

16. Goth, R.W., and Webb, R.E. 1981. Sources and genetics of host resistance in vegetable crops. Pages 377-411 in: Fungal Wilt Diseases of Plants. M.E. Mace, A.A. Bell and C. H. Beckman, eds. Academic Press, NY.

17. Green, R.J., Jr. 1981. An overview. Pages 1-24 in: Fungal Wilt Diseases of Plants. M. E. Mace, A.A. Bell and C.H. Beckman, eds. Academic Press, NY.

18. Hafez, A.A.R., Stout, P.R., and DeVay, J.E. 1975. Potassium uptake by cotton in relation to Verticillium wilt. Agron. J. 67:359-361.

19. Hancock, J.G., and Huisman, O.C. 1981. Nutrient movement in host-pathogen systems. Ann. Rev. Phytopathol. 19:309-331.

20. Heale, J.B., and Isaac, I. 1963. Wilt of lucerene caused by species of *Verticillium*. IV. Pathogenicity of *V. albo-atrum* and *V. dahliae* tto lucerne and other crops; spread and survival of *V. albo-atrum* in soil and in weeds; effect upon lucerne production. Ann. Appl. Biol. 52:439-451.

21. Howard, R.J. 1985. Local and long-distance spread of *Verticillium* species causing wilt of alfalfa. Can. J. Plant Pathol. 7:199-202.

22. Huber, D.M. 1980. The role of mineral nutrition in defense. Pages 381-406 in: Plant Disease, An Advanced Treatise. Vol. 5. J. Horsfall and E. Cowling, eds. Academic Press, NY.

23. Huber, D.M., and Watson, R.D. 1970. Effect of organic amendment on soil-borne plant pathogens. Phytopathology 60:22-26.

24. Huber, D.M., and Watson, R.D. 1974. Nitrogen form and plant disease. Ann. Rev. Phytopathol. 12:139-165.

25. Huisman, O.C., and Ashworth, L.C., Jr. 1976. Influence of crop rotation and survival of *Verticillium albo-atrum* in soils. Phytopathology 66:978-981.

26. Isaac, I. 1956. Some soil factors affecting Verticillium wilt of Antirrhinum. Ann. Appl. Biol. 44:105-112.

27. Isaac, I. 1957. The effects of nitrogen supply upon the Verticillium wilt of Antirrhinum. Ann. Appl. Biol. 45:512-515.

28. Isaac, I. 1957. Wilt of lucerne caused by species of *Verticillium*. Ann. Appl. Biol. 45:550-558.

29. Isaac, I., and Heale, J.B. 1961. Wilt of lucerne caused by species of *Verticillium*. III. Viability of *V. albo-atrum* carried with lucerne seed; effects of seed dressings and fumigants. Ann. Appl. Biol. 49:675-691.

30. Jones, J.P., and Overman, A.J. 1986. Verticillium wilt (race 2) of tomato. Plant Pathol. Circ. #284. Fla. Dept. Agric. & Consumer Serv. Div. of Plant Industry.

31. Jordan, V.W.L., Sneh, B., and Eddy, B.P. 1972. Influence of organic soil amendments on *Verticillium dahliae* and on the microbial composition of the strawberry rhizosphere. Ann. Appl. Biol. 70:139-148.

32. Keinath, A.P., and Millar, R.L. 1986. Persistance of an alfalfa strain of *Verticillium albo-atrum* in soil. Phytopathology 76:576-581.

33. Keyworth, W.G. 1942. Verticillium wilt of the hop (*Humulus lupulus*). Ann. Appl. Biol. 29:346-357.

34. Keyworth, W.G., and Hewitt, E.J. 1948. Verticillium wilt of the hop (*Humulus lupulus*). V. The influence of nutrition on the reaction of the hop plant to infection with *Verticillium albo-atrum*. J. Hort. Sci. 24:219-227.

35. Marschner, H. 1986. Mineral nutrition of higher plants. Academic Press, Orlando, FL. 674 p.

36. Panton, C.A. 1964. A review of some aspects of the wilt pathogen *Verticillium albo-atrum*, Rke. et Berth. Acta Agric. Scand. 14:97-112.

37. Panton, C.A. 1967. The breeding of lucerne, *Medicago sativa* L. for resistance to *Verticilllium albo-atrum* Rke. et Berth. II. The quantitative nature of the genetic mechanism controlling resistance in inbred and hybrid generations. Acta Agric. Scand. 17:43-52.

38. Pegg, G.F. 1985. Life in a blackhole--the microenvironment of the vascular pathogen. Trans. Brit. Mycol. Soc. 85:1-20.

39. Ranney, C.D. 1962. Effects of nitrogen source and rate on the development of Verticillium wilt of cotton. Phytopathology 52:38-41.

40. Roberts, F.M. 1943. Factors influencing infection of the tomato by *Verticillium albo-atrum*. Ann. Appl. Biol. 30:327-331.

41. Roberts, F.M. 1944. Factors influencing infection of the tomato by *Verticillium albo-atrum*. II. Ann. Appl. Biol. 31:191-193.

42. Schnathorst, W. C. 1981. Life cycle and epidemiology of *Verticillium*. Pages 81-111 in: Fungal Wilt Diseases of Plants. M.E. Mace, A.A. Bell and C.H. Beckman, eds. Academic Press, NY.

43. Selvaraj, J.C. 1974. Alteration of the production and activity of pectinases of *Verticillium dahliae* by calcium chloride and sodium chloride. Indian Phytopathol. 27:437-441.

44. Sewell, G.W.F., and Wilson, J.F. 1967. Verticillium wilt of the hop: field studies on wilt in a resistant cultivar in relation to nitrogen fertilizer applications. Ann. Appl. Biol. 59:265-273.

45. Sivaprakasam, K., and Rajagopalan, C.K.S. 1974. Effect of nitrogen on the incidence of Verticillium wilt disease of egg plant caused by *Verticillium dahliae* Kleb. Plant and Soil 40:217-220.

46. Talboys, P.W. 1987. Verticillium wilt in English hops: retrospect and prospect. Can. J. Plant Pathol. 9:68-77.

47. Wilhelm, S. 1951. Effect of various soil amendments on the inoculum potential of the Verticillium wilt fungus. Phytopathology 41:684-690.

48. Wilhelm, S. 1975. Sources and nature of Verticillium wilt resistance in some major crops. Pages 166-171 in: Biology and Control of Soil-borne Plant Pathogens. G.W. Bruehl, ed. American Phytopathological Society Press, St. Paul, MN.

49. Wilhelm, S. 1981. Sources and genetics of host resistance in field and fruit crops. Pages 300-376 in: Fungal Wilt Diseases of Plants. M.E. Mace, A.A. Bell and C.H. Beckman, eds. Academic Press.

50. Wilhelm, S., and Paulus, A.O. 1980. How soil fumigation benefits the California Strawberry Industry. Plant Dis. 64:264-270.

51. Wilhelm, S., Sagen, J.E., and Tietz, H. 1985. Phenotype modification in cotton for control of Verticillium wilt through dense plant population culture. Plant Dis. 69:283-288.

THE ROLE OF NUTRITION IN THE TAKE-ALL DISEASE OF WHEAT AND OTHER SMALL GRAINS

Don M. Huber
Purdue University
West Lafayette, Indiana 47907

Few diseases respond as dramatically to nutrition as take-all of small grains caused by *Gaeumannomyces graminis* (Sacc.) von Arx & Olivier (*Ophiobolus graminis*). The response to specific nutrients and cultural practices such as pH, crop sequence, and tillage, which condition the abundance, form, or availability of nutrients is so pronounced as to provide effective levels of disease control when combined with other management practices (26,42). Losses from take-all are generally severe when plants are deficient in any of the essential mineral elements (43,68). There is an inverse relationship between the nutritional status of the plant and the severity of take-all.

Twelve of the thirteen principal mineral nutrients required for plant growth are reported to affect take-all either individually or collectively. Eleven elements (N, P, K, Ca, Cl, Cu, Fe, Mg, Mn, S, Zn) are reported to reduce take-all severity while six (N, P, Ca, K, Mg, Mo) are also reported to increase disease severity (Table 1) (32,41-42,43,74-75). The reported effect of a nutrient on take-all may be determined by the form or application rate of the nutrient, through a commonly associated ion, or relative to the nutritional status of the plant. Frequent failure to report these variable or environmental conditions sometimes makes evaluation of divergent results difficult. Understanding the individual and interacting effects of these nutrients in a crop production system can provide a basis for the cultural control of take-all.

Table 1. Mineral elements affecting take-all of cereals.

Increase Take-all	Reduce Take-all
Potassium nitrate	Potassium chloride
Phosphorus excess	Phosphorus sufficiency
Calcium carbonate (lime)	Sulfur
Magnesium carbonate	Magnesium chloride
Magnesium sulfate	Calcium chloride
Molybdenum	Manganese
	Iron
	Zinc
	Copper chloride

NITROGEN

The nearly universal deficiency of N in cultivated soils for optimum cereal growth probably accounts for the high frequency of reports associating a shortage of N with increased take-all. Nitrogen deficiency was reported as a predisposing factor for take-all in the 1900's and has since been associated commonly with severe outbreaks of take-all in all cereal-growing countries (3,11,25,39,40,43,68,77). In a deficiency situation, disease severity declines as the rate of applied N increases (8,17,20-23,26,39,41-43,52,82,88,90). As N rates exceed the optimum required for plant growth, variable effects on take-all have been reported (13,38,52). Nitrogen may have little effect on mild infections of take-all (85), but may have a pronounced effect in maintaining yield potential when more severe disease occurs even though the amount of infection may not be reduced (44,52,87). This increased tolerance to disease has been attributed to enhanced plant vigor, production of more roots, and greater availability of N to offset the root rot and reduced root absorption efficiency caused by the pathogen (22,32,42,44).

Conflicting reports of the effect of N on take-all frequently can be attributed to failure to consider the many interacting factors which influence N availability and the mechanisms involved in control (12,53). Nitrogen, as the fourth most abundant element in plants, is available through biological mineralization of complex soil organic matter, microbial fixation of atmospheric N, or as organic and inorganic fertilizer amendments. Mineralization and biological fixation are dynamic processes providing N initially in the ammoniacal (NH_4) or reduced form. The subsequent biological oxidation of NH_4-N to nitrate (NO_3-N) results in the availability of several forms of N throughout plant growth (Fig. 1) (54,55). Unlike the positively charged NH_4 ion, the negatively charged NO_3 ion is freely mobile in the soil solution and subject to leaching as well as denitrification losses which predispose wheat to take-all (55,56).

Although wheat, like other plants, uses either form of N for growth, plant composition, microbial activity in the rhizosphere, availability of minor elements, and the interactions of factors which influence disease severity are markedly influenced by the form of N utilized. Amine, amide, and protein composition are higher with NH_4-N than NO_3-N (55) and NH_4-N is more "metabolically efficient" so that the demand placed on reserve carbohydrates is lower than observed with NO_3-N (35,97). Absorbed NH_4-N is rapidly conjugated with carbohydrates in the root system to form amino acids and to prevent toxicity. This metabolic activity creates a dynamic carbon sink in the root system which may increase photosynthetic efficiency as accumulated sugars are translocated out of leaf tissues (47,97). Modified microbial activity in the rhizosphere is associated with the increased root metabolism and diversity of nutrients in root exudates following absorption of NH_4-N (47,50). Increased availability and uptake of Fe, Mn, P, and Zn which reduce take-all severity are associated with the modified rhizosphere microbial activity and decrease in rhizosphere pH induced by NH_4 uptake (5,10,32,36,37,47,50,75,76,90). Unlike NH_4, NO_3-N is readily transported as the NO_3 ion and stored until metabolically reduced to useable forms of N in leaf and stem tissues. Severe take-all of NO_3-N-fertilized wheat under drought or high temperature stress conditions which prevent NO_3 utilization (35) is probably a reflection of N stress through the plant's inability to utilize the NO_3-N present in tissues. Leaching and denitrification losses of NO_3-N also may contribute to the reduced effectiveness of this form of N in reducing take-all (42,43,55).

Thus, the form of N is of even greater importance than the rate of N in reducing take-all (39,40,42,52,53,64,88,95). Much of the early confusion regarding the effect of N on take-all can be explained by the divergent effects of different forms of N and different environmental and cultural practices in their application and use (43,52-55,63,90). Previous crop, soil pH, moisture, temperature, tillage, weed competition, plant density, stage of plant growth, and availability of other nutrients and carbohydrates influence the stability of a particular form of N in soil (40,42,52-55). Ammoniacal sources of N, i.e.

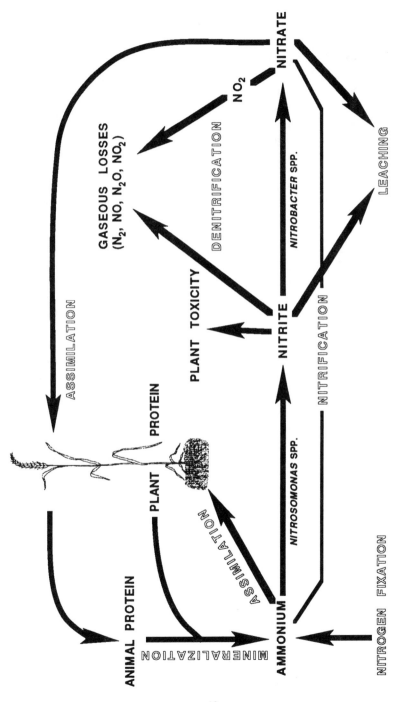

FIGURE 1. SCHEMATIC OF THE NITROGEN CYCLE

49

NH_3, NH_4OH (aqua ammonia), $(NH_4)_2SO_4$, NH_4Cl, $(NH_4)_2PO_4$, and urea reduce take-all (6,37,40,42,53,90) while NO_3 sources of N, i.e. $Ca(NO_3)_2$, KNO_3, urea-NH_4NO_3 solutions (UAN), and NH_4NO_3 increase infection. Yield losses are generally reduced with either form of N (21,40,42,44,53) since both forms increase the availability of N to compensate for the loss of absorption in diseased tissues.

Although both forms of N are used by the wheat plant, in the absence of conditions inhibiting nitrification, a preponderance of NO_3-N may be available to the plant and pathogen (55) and the rate of nitrification may influence the intensity of effect observed with different sources of N (40). Ammoniacal N may be relatively stable in acid soils, with certain crops in the rotation, or with chemicals which inhibit nitrification, but relatively unstable (rapidly nitrified) in neutral to alkaline soils or following crops or manures which stimulate nitrification (7,43,44,53-55,60). Control of take-all with NH_4N is generally enhanced under conditions which restrict nitrification (42-44,53-55,86,88,90). Specific biocides such as nitrapyrin [2-chloro-6-(trichloromethyl) pyridine; N-Serve, Dow Chemical, U.S.A.] and etradiazol [5-ethoxy-3-trichloromethyl-1,2,4-thiadiazole; Dwell, UniRoyal Chemical Co.], etc., which inhibit nitrification, not only ensure the availability of N for crop growth and yield later in the season, but also provide an effective means of reducing take-all when applied with NH_4-N by maintaining an optimum 3:1 $NH_4:NO_3$ ratio for disease suppression (Table 2) (7,44,53,55,56,88,90). Much of the effect of Cl fertilizers in reducing take-all has been correlated with the suppression of nitrification (7,76).

Manures which stimulate nitrification are generally not as effective in reducing take-all as commercial fertilizers (44,77). The variable effects of organic amendment with manures, blood-meal, green manure crops, or crop rotation on take-all depend on the sufficiency of N or P supplied and their effect on nitrification (3,11,17,20,23,28,42,52,55,60,93). Although N liberated from organic manures may increase the survival of *G. graminis* in crop residues, it also increases the vigor and tillering capacity of the plant to provide a net beneficial effect of increased grain yield (18-20,23). If subterranean clover or alfalfa (lucern), which enhance nitrification, are used to increase soil fertility preceding wheat in the rotation, take-all may increase in severity (3,11,42,52,53,60), while a preceding crop of lupins, which inhibit nitrification, may reduce take-all even though the soil N content is similar for both crops (52,60).

Table 2. Effect of nitrapyrin applied with ammoniacal nitrogen on grain yield and take-all of winter wheat in the Eastern Corn Belt of the United States, 1985.[a]

Nitrogen	Sand		Sandy Loam		Silt Loam	
Rate	Yld	Dis.	Yld	Dis.	Yld	Dis.
kg/ha	kg/ha	WH[b]	kg/ha	WH	kg/ha	WH
0	1270	30	938	50	2410	8
45	1540	26	2350	23	2950	3
45+ NS[c]	1876	16	3080	11	369	2
90	1680	18	3150	19	4020	1
90 + NS	2750	9	4220	7	4420	1
135	2550	12	3220	20	4420	1
135+ NS	3080	3	4760	6	4560	1

[a] All fields are the third consecutive wheat crops with severe natural infections of *G. graminis*.

[b] WH = % white (sterile) heads

[c] NS = Nitrapyrin at 0.55 kg/ha

Acid soils generally have low rates of nitrification. Liming soil to neutralize soil acidity increases nitrification and take-all. Symptoms of take-all sometimes have been confused with Mg deficiency resulting in recommending the application of dolomitic lime (Ca, $MgCO_3$). This has generally increased disease severity and reduced the availability of Mn and other minor elements. Application of high rates of chlorides such as KCl or NaCl to seed along with NH_4 sources of N, especially NH_4Cl, reduces the severity of take-all by inhibiting nitrification (5,7,36,37,47,72,76) and possibly by reducing NO_3-N uptake while increasing the uptake of other ions (57).

The effect of time of N application on take-all is related to the form of N available or to N stress before N fertilization. The common practice of split (multiple) applications of N reflects the effects of nitrogen loss from leaching or denitrification. Fall applications of low rates of N preplant for winter wheat, followed by early spring top-dressings, may be adequate for "late" seeded wheat or on soils with high levels of residual N. This may result in N deficiency and predisposition to take-all when fall conditions permit excessive growth which exhausts the readily available residual N. Delayed top-dressing of N in the spring because of inclement weather may also result in N stress of the rapidly growing wheat. Thus, application of N to coarse-textured, sandy soil after most of the heavy spring rains are past has provided better control of take-all and higher grain yields than early application of non-stabilized N which is subjected to winter or spring leaching (43,55,56,83,85,86). An economically effective alternative to multiple applications of N for winter wheat in the midwestern U.S. has been the preplant application of nitrapyrin stabilized NH_3 (44,56) which has also reduced the severity of sharp eye-spot, Rhizoctonia spring blight, and Pseudocercosporella foot rot (40,42,44).

In contrast, nitrification of fall-applied N for spring wheat increases take-all, while spring application of NH_4-N markedly suppresses the disease and increases grain yield. An additional application of NH_4-N in the spring partially offsets the increased disease observed with the fall- nitrified N (Table 3) (52).

Table 3. Effect of time of application of ammonium nitrogen on take-all and yield of irrigated spring wheat (43).

Nitrogen[a] Applied	Rate	Take-all Index[b]	Grain Yield
kg/ha			kg/ha
0	0	1.9	2610
Fall	83	2.8	1740
Spring	83	0.1	5290
Fall + Spring	83 + 28	1.9	2350

[a] Broadcast, incorporated as NH_4SO_4
[b] Root rot index of 0-4 where 0 = no infection; 4 = 100% infection and non-functional root system.

Take-all also may be increased during mineralization of carbonaceous organic residues which immobilize N even though N is eventually available for crop growth. The effect of immobilizing N on take-all also has been demonstrated by comparing gelatin as a readily available N source which reduces take-all with starch and glucose which immobilize N during stimulated microbial activity and increase take-all (58).

The effect of N on take-all is interrelated with the availability of other nutrients and, in certain marginal soils, high levels of N may predispose plants to micronutrient deficiencies not apparent at lower levels of N (32). Copper, B, and Mn interact in N metabolism and may exert their effect on take-all through their role in N metabolism. The different forms of N may influence disease severity through their differential metabolic activity, effect on rhizosphere microorganisms, or uptake of minor elements. The metabolism of NH_4-N to amino acids in the root system and their subsequent translocation throughout the plant create a dynamic carbon sink in the roots as photosynthate (especially organic acids) is translocated downward to be conjugated with the absorbed NH_4-N (Fig. 2). This induced carbon sink removes feedback inhibition of photosynthesis (65), provides energy for root defenses, decreases rhizosphere pH (7,10,76,88,89,90) and increases the nutrients in root exudates. Increased root exudation selectively stimulates rhizosphere microorganisms and suppresses Mn oxidizers (44,50). Thus, the availability of Fe, Mn, and Zn are enhanced by NH_4 which also increases P uptake (32,75,98). These effects will be discussed further with the specific elements.

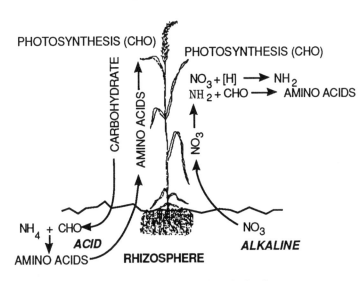

FIGURE 2. Schematic presentation of carbohydrate movement and rhizosphere environment with absorption of N forms.

There is a good correlation between the effect of most conditions affecting take-all and their effect on nitrification or N availability (Table 4); however, it should not be assumed that this is the only effect of these conditions influencing disease severity. Rather, these effects should be considered synergistic with nutrition. The association of severe take-all with high rainfall, irrigation, or waterlogged soil

Table 4. Correlation of factors influencing the form of nitrogen in soil with the severity of take-all.

Factor	Effect on	
	Nitrification	Disease
Nitrate nitrogen	-	Increase
Ammoniacal nitrogen	-	Decrease
Liming	Increase	Increase
Acid Soils	Decrease	Decrease
Manure	Increase	Increase
Nitrificides	Decrease	Decrease
No-till tillage	Increase	Increase
Chloride	Decrease	Decrease
Alfalfa precrops	Increase	Increase
Lupine precrops	Decrease	Decrease

conditions may reflect increased N loss from leaching or denitrification (3,11,22,26,52,66,77). Waterlogged soils subject to denitrification are especially conducive to take-all (55,77). Reduced denitrification may account for the reduced loss from take-all after soil drainage (16).

Tillage, which leaves a loose, cloddy seedbed, or ploughing under crop residues which leaves the top soil open and highly aerobic, results in a preponderance of NO_3-N and increases the severity of take-all (43, 55, 84).

A major effect of weeds in predisposing wheat to take-all is through competition for nutrients. We have observed a greater predisposition to take-all with cheat grass (*Bromus secalinus* L.) as an early competitor for nutrients than with the later-developing quackgrass (*Agropyron repens* (L.) Beauv.) although both increase take-all. Where weed competition for nutrients is intense, herbicides and some soil fumigants may indirectly reduce take-all by reducing this source of competition for nutrients (68). The recommendation to reduce the seeding rate where take-all is a problem (3,28) increases the availability of nutrients on a per plant basis and reduces competition for limited nutrients in soil (8,24,25,27,83). Early seeding of winter wheat may exhaust limited nutrient reserves before dormancy and predispose plants to take-all. Increased prevalence of take-all on coarse-textured soils (3,21,29,43,68) compared to fine-textured soils, also is generally related to their over-all nutrient status.

PHOSPHORUS

Phosphorus, like N, is commonly deficient in wheat soils and PO_4 fertilizers are widely used in combination with N to reduce the severity of take-all. Phosphate fertilization, even in excess of a sufficiency for crop growth, has been effective in reducing losses from take-all (3,93) and large, single applications of P may decrease take-all more effectively than smaller, annual dressings (65) as long as other nutrients are not deficient. The beneficial effects of P in reducing take-all are a reflection of stimulated root development (3) and increased host resistance (3,31,94). Take-all occurs in the presence of vesicular-arbuscular mycorrhizae which increase the P status of plants. Reduced disease was related to decreased root exudation of amino acids and carbohydrates which suppressed pathogen activity rather than to increased root growth (31). Phosphate fertilizers are less effective in reducing take-all where N deficiency is limiting crop growth (3,14,93,94) and disease severity may be increased by P under severe N (94) or Mn deficiency (Table 5). High levels of residual P in heavily manured soils have predisposed plants to take-all because of increased microbial activity and complexing with Mn to reduce availability of

Table 5. Effect of nitrogen and phosphorus on take-all of winter wheat in a soil deficient for manganese.[a]

Nitrogen	Phosphorus	Take-all	Grain Yld
kg/ha	kg/ha	WH[b]	kg/ha
0	0	1	1200
0	110	4	2080
55	0	3	3350
55	110	9	3550
110	0	1	3550
110	110	10	4220

[a] Silty clay loam soil, third consecutive wheat crop. Southeastern Indiana.

this essential element (57). The best disease control is obtained with the application of both N and P (92,93). Combinations of P and K also reduce take-all (3,22,23,30), although the greatest disease reduction is generally obtained with a well-balanced fertilizer including N (21,23,32,44). The availability of insoluble soil P is primarily dependent on microbial activity in the rhizosphere, the solvent activity of root exudates, and rhizosphere pH, which may explain why P uptake is enhanced by NH_4-N (42,98). Phosphate fertilizers are less effective in reducing take-all where N deficiency is limiting crop growth (3,14,93,94) and disease severity may be increased by P under severe N (94) or Mn deficiency (Table 5). High levels of residual P in heavily manured soils have predisposed plants to take-all because of increased microbial activity and complexing with Mn to reduce availability of this essential element (57). The best disease control is obtained with the application of both N and P (92,93). Combinations of P and K also reduce take-all (3,22,23,30), although the greatest disease reduction is generally obtained with a well-balanced fertilizer including N (21,23,32,44). The availability of insoluble soil P is primarily dependent on microbial activity in the rhizosphere, the solvent activity of root exudates, and rhizosphere pH, which may explain why P uptake is enhanced by NH_4-N (42,98).

POTASSIUM

Few references imply a separate role for K relative to take-all and it is important to maintain a balanced fertility program as previously discussed. Potassium is considered a mobile regulator of cellular activity and the level of K in the plant depends on the availability of Mg and Ca, which are, in turn, influenced by pH. Potassium availability is enhanced by Ca in neutral but not in acid soils, and a deficiency of K impairs P utilization just as though a N deficiency exists (47). Take-all is increased by K if N and P are deficient (71,92), but decreased by K when applied with adequate N and P (3,22,23,30,44,72,74).

The associated anion applied with potassium may have an independent effect on disease that is attributed to K. This effect is especially pronounced with KCl, where the reduction in take-all (36,37,72,87) observed with KCl can also be achieved with NH_4Cl but not K_2SO_4. The chloride ion inhibits nitrification, is a competitive inhibitor of NO_3-N uptake and reduces take-all when applied with $(NH_4)_2SO_4$ or NH_4Cl, but not when applied with NO_3 sources of N

(7,42,47,76). Potassium nitrate increases take-all severity even though yields may be increased (47).

CALCIUM

Calcium, commonly applied as the carbonate salt or CaO to neutralize soil pH, also is used as a fertilizer to supply N [$Ca(NO_3)_2$] and S ($CaSO_4$, gypsum). Calcium has a critical metabolic role in carbohydrate removal, neutralization of cell acids, cell wall deposition, and formation of pectates in the middle lamella. The dual effects of lime in providing Ca for nutrition and in increasing soil pH, with its multiple effects on minor nutrient availability, have not always been separated.

Numerous investigators report increased take-all after applying lime. Take-all has been recognized as a serious disease of wheat on calcareous soils for many years (42,43). Rosen and Elliott (77) reported that take-all became so severe after liming that the plots were not harvested, and Ophiobolus patch disease of turf is especially severe along limed areas of cricket outfields (91). Wheat is severely damaged by take-all following soybeans on soils limed to bring the pH up to 6.0 to favor nodulation of soybeans by *Rhizobium* (75,80). In pot culture (using NO_3-N in a $Ca(NO_3)_2/KNO_3$ mixture) with a range of soil pH from 4.5 to 8.5, Reis, et al. (75) reported that take-all increased as the pH increased from 4.5 to 8.5 at their lowest level of additional Ca (as $CaSO_4$). They also reported that take-all increased at pH's below 6.5 as the level of Ca increased. Interactions of pH and level of Ca were observed also at the highest pH tested (8.5) where the intermediate level of Ca markedly reduced take-all to levels observed at the lower pH. Although tissue levels of Cu, Fe, Mn, and Zn were lower at the two highest pH levels tested, little effect on their uptake was reported at pH's below 7.5 except for Fe which was higher at pH's 5.5 and 6.5 than 4.5. There was no specific correlation of these levels of minor elements with disease severity. In a companion study where the level of Mg (as $MgSO_4$) rather than Ca was varied over the same pH range, little effect of pH or added Mg was observed (75). Modification of the pH effect on take-all by the addition of $MgSO_4$ and increased take-all at low soil pH's by increased levels of $CaSO_4$, indicate Ca and pH may be increasing the availability of Mg and/or Mo (44) which increase take-all. Elimination of the effect of pH on take-all by Mg (75) may imply a direct role of Ca in root tissues on susceptibility to take-all since the Ca content of plant tissues is generally reduced by Mg (42).

Calcium reduces the adverse effects of excessive levels of P and other elements on Mn availability, and liming may reduce take-all under these conditions even though it may increase the availability of Mg and Mo which can increase take-all. Like K, the ion associated with Ca may be the predominant influence on take-all. Thus, $Ca(NO_3)_2$ and $CaCO_3$ both increase take-all (43,44,75) while $CaCl_2$ may reduce take-all when applied with NH_4-N.

SULFUR

A deficiency of·S, like other mineral deficiencies, increases the severity of take-all (42,74). The limited number of observations of S and take-all probably reflects the sufficient availability of this mineral element in most soils. Sulfur deficiency is rare in industrialized countries because adequate levels of various sulfur oxides are present for optimum plant growth. Sulfur is reduced in the plant (similar to NO_3-N) and is incorporated into amino acids, proteins, vitamins, aromatic oils, and ferredoxins. When needed, it is commonly applied as gypsum ($CaSO_4$), but wettable S or sulfuric acid have sometimes been used (41).

Elemental S has generally been thought to affect disease severity through its acidifying effect on oxidation (42,43). Ammonium sulfate has been a preferred source of N under take-all conditions as much because of the NH_4-N as the SO_4-S (68). Davidson & Goss (14) reported a significant reduction of take-all after application of wettable S alone or in combination with urea and P. The S and urea combination provided more effective control of take-all than the combination of S and P. Since take-all was not affected by sulfuric acid sprays applied in the fall and spring to control Pseudocercosporella foot rot (24,25), possible pH effects should not necessarily detract from consideration of the vital role of S as a mineral element with its metabolic ramifications.

CHLORIDE

Chloride, applied at relatively high rates as a salt of NH_4, Ca, K, or Na, reduces take-all when applied with NH_4-N but not NO_3 (6,7,36,37,47,72,76). Application of chloride fertilizers to fields with a history of take-all has become common practice in parts of the Pacific Northwest (7). It is generally band-applied with P and NH_4-N. Suppression of take-all is maximized when Cl is applied in conjunction with other management practices that reduce disease severity such as delayed seeding (7,95). The mechanism of control with Cl is not known; however, Cl apparently does not act as a micronutrient since large amounts are required. Possible mechanisms for reducing take-all are by suppressing nitrification and NO_3-N uptake, or by influencing availability or function of Mn (73) or some other mineral element as discussed later in this chapter (6,7,41,42,47,53).

MAGNESIUM

Magnesium tends to be deficient in acid soils because of limited solubility; high levels of K or Ca may inhibit uptake of Mg (44). The depletion of Mg from soil observed with continued cereal cropping or the use of $(NH_4)_2SO_4$ fertilizers (1) may be an important factor for the take-all decline phenomenon; however, plants growing in nutrient conditions most conducive to take-all have low Mg levels in tissues (38). Deficiency symptoms are readily cured by the addition of dolomitic lime or other sources of Mg. Increased severity of take-all after application of dolomitic lime to neutralize soil acidity is frequently observed; however, Reis, et al. (75) found little effect of pH on take-all following the addition of relatively high levels of Mg in pot culture.

Greenhouse-grown plants deficient in Mg are severely damaged by take-all, and disease severity is reduced by adding Mg (38,74). Magnesium, under these conditions, is most effective in reducing take-all when used in combination with a mixture of NH_4-N and NO_3-N and least effective when added with NH_4-N alone as a possible result of ion competition from nutrient solution (38,42). Co-injection of 0.6 to 1.0 kg/ha chelated Mg with nitrapyrin-stabilized NH_3 (44) into field soils severely deficient in Mg provided a mixture of both forms of N for plant uptake (NH_4-N from applied N and NO_3-N from non-stabilized, mineralized N). This method also reduced take-all severity and increased grain yield of take-all tolerant wheat, but had little effect on a susceptible wheat variety (Table 6). Similar applications of Mg to soils deficient in Mn resulted in extremely severe take-all where plants died before heading.

A direct role of Mg in pathogenesis is indicated since infected roots maintain normal levels of Mg (1), root exudates of take-all resistant oats contain avenacin which inhibits the utilization of Mg by *G. graminis* (69), and Mg uptake is reduced in the presence of NH_4-N (38). Inhibition of Mg uptake by high levels of K (44) may partially explain the reduction in take-all following application of KCl in combination with NH_4-N (5,36,37,72).

Table 6. Yield of take-all susceptible and tolerant winter wheat fertilized with coinjected[a] minor elements and anhydrous ammonia under severe take-all pressure in the field.

Minor Element[b]	Variety	
	Auburn	Beau
	---kg/ha---	
0	2950	2750
Copper	2750	2880
Iron	2950	2880
Magnesium	3350	2880
Manganese	3750	3350
Zinc	2950	2880

[a] Coinjected with 90 kg N/ha on 30 cm with 0.55 kg/ha nitrapyrin through flexible NH_3 tines (Harlan Manufacturing Co., Harlan, IA) to a low Mg, sandy loam, field soil cropped to wheat the previous two years.

[b] Minor elements were injected into the flowing NH_3 as a 1.0 kg/ha aqueous solution in 24 l/ha with a DecaH (DecaH Manufacturing, West Point, IN) Ad-A-Matic down-stream injection system.

MOLYBDENUM

Molybdenum, coinjected with NH_4-N, has increased take-all at two locations in Indiana, presumably by interfering with Mn uptake. Very high levels of P enhance Mo uptake (44) and increase take-all. Take-all increased following application of Mo and other minor elements without N, but was reduced when Mo was applied with $(NH_4)_4SO_4$ (28).

COPPER

Reduction in take-all by high rates of copper (74) may result from a fungistatic effect (32); however, the role of Cu in lignification (2), its interaction with N, its increased availability in acid soils as CuS, and the importance of lignitubers in delaying pathogenesis indicate a potential direct metabolic role of Cu in the plant's defense against *G. graminis*. Either form of N or P may aggravate deficiencies of Cu and Zn because of the high affinity of these elements for proteins (32), while only NO_3-N increases the severity of Mn deficiency (96). This differential effect of N forms and the often favorable effect of liming on increasing Cu availability, indicate that Cu is important in take-all primarily under deficiency conditions.

MANGANESE

Manganese interacts with N metabolism and is intimately involved in carbohydrate synthesis, photosynthesis and the synthesis of phenols and other compounds associated with the defense of plants against pathogens.   In addition, root development is positively correlated with the Mn content of soil (62). Until recently, there were few reports of an association of Mn with take-all, perhaps because of difficulty in establishing a deficiency under pot culture conditions (32,59). Deficiency in the field also may be ephemeral depending on seasonal, soil and biological conditions (32,44). Manganese is readily available in acid soils with negligible take-all and is less available as soils become highly conducive to take-all upon liming (32,42,43,74,75).

Our research indicates an interacting and predominant role of Mn with other elements in the control of take-all. Although recognizing that a deficiency in any nutrient will predispose cereals to take-all, the importance of Mn availability became apparent to us through a long-term field study of take-all decline in which we were assaying for soil physical, chemical and biological factors which might relate to disease suppression or enhancement. Detailed assay of several areas severely damaged by take-all failed to indicate any significant differences from adjacent areas of the field where take-all was relatively mild; yet, these severely diseased areas were relatively stable year after year. Soybeans planted in one of these study sites on sandy loam soil at pH 6.0 exhibited severe Mn deficiency symptoms only on the severe take-all areas. Although soil levels of Mn were marginal, previous soil analyses indicated there were no significant differences between the severe and adjacent mild take-all areas. Since Mn availability for plant uptake is influenced not only by soil chemical reactions (in the available, reduced form below pH 5.2 and in the non-available, oxidized form above 7.8), but also by biological activity between these two extremes of pH 5.2 and 7.8 (44), we assayed for Mn oxidizing populations. The population of Mn oxidizers in soybean rhizospheres in the severe take-all areas was several times higher than in adjacent, mild take-all areas. Plant tissue analyses corroborated the effects of rhizosphere biological activity on Mn availability (50). These observations were confirmed at various locations where the population of Mn oxidizing bacteria in the rhizospheres of severely infected wheat plants was compared with adjacent, mildly diseased plants. These results indicated that the biological component determining Mn availability was a critical, and perhaps predominant, factor in determining Mn amelioration of take-all severity except under highly calcareous soils or the most marginal soil conditions in which the levels of Mn were deficient regardless of the soil redox potential or biological activity. This can be schematically presented as:

pH:	Acid, <5.5	Biological	>7.8, Alkaline
Mn form:	Mn^{2+}		Mn^{4+}
Availability:	Available		Non-available

In comparison, most conditions influencing take-all severity have a correlative influence on the Mn oxidizing bacteria and Mn availability (Table 7). The increased uptake of Mn with NH_4-N fertilizers is associated with a reduction in the population and activity of Mn oxidizers in the rhizosphere (50); oxidizing bacteria are reduced when nitrification is inhibited (47,50).

Wheat varieties such as "Auburn," which are tolerant to take-all, have a much lower population of Mn oxidizers in their rhizospheres and higher tissue levels of Mn than susceptible cultivars such as "Beau" or highly susceptible breeding lines we have evaluated (Table 8) (50). Bacterization with an oxidizing bacterium increased take-all. Similarly, seed treatments with peat, which immobilize Mn, increased take-all and nullified the beneficial effects of seed bacterization with plant growth promoting rhizobacteria (46,50,51,78).

Oats as a precrop to wheat reduces take-all and markedly suppresses the population of Mn oxidizers, while a soybean precrop frequently enhances take-all and the population of Mn oxidizers in subsequent wheat rhizospheres. Older oat varieties which were highly sensitive to gray-speck disease, i.e. Mn deficiency, had high populations of oxidizers in the rhizospheres (41). Current varieties resistant to gray-speck disease suppress Mn oxidizers in the rhizosphere and also are tolerant to take-all.

Table 7. Correlation of conditions affecting take-all with the availability of manganese.

Condition	Effect on:	
	Take-all	Mn Availability
Liming	Increase	Decrease
Nitrate nitrogen	Increase	Decrease
"Short" monocropping	Increase	Decrease
Plant stress	Increase	Decrease
Manuring	Increase	Decrease
Loose seedbed	Increase	Decrease
Soybean precrop	Increase	Decrease
Alfalfa precrop	Increase	Decrease
High moisture	Increase	Decrease
Alkaline pH soils	Increase	Decrease
Heavy seeding	Increase	Decrease
Ammonium nitrate	Decrease	Increase
Tolerant cultivars	Decrease	Increase
Nitrification inhibitors	Decrease	Increase
Acid pH soils	Decrease	Increase
Oats precrop	Decrease	Increase
Late seeding	Decrease	Increase
Manganese fertilization	Decrease	Increase
Chloride	Decrease	Increase

Liming and NO_3-N increase the population of Mn oxidizers and increase take-all. Ammoniacal N provides an acidic environment, suppresses Mn oxidizers in the rhizosphere, and reduces take-all. Wet spring conditions conducive to severe take-all increase the activity of Mn oxidizers and facilitate the loss of residual N by leaching and denitrification. Plants under stress frequently have higher populations of Mn oxidizers and are more susceptible to take-all than non-stressed plants. Other cultural and environmental conditions influencing take-all severity are negatively correlated with their effects on the biologically-induced deficiency or availability of Mn (Table 7).

Table 8. Tissue manganese in wheat cultivars tolerant and susceptible to take-all after fertilization with ammoniacal nitrogen and inhibiting nitrification.

Treatment		Cultivar	
N[a]	Nitrapyrin	Auburn	Beau
----------kg/ha---------		μg Mn/g tissue[c]	
0	0	8	11
65	0	14	11
65	0.55	18	15

[a] N, as NH_3, applied to a sandy field soil low in Mn and cropped to wheat the two previous years.
[b] Auburn is tolerant to take-all; Beau is susceptible.
[c] 20 μg/g is considered minimum sufficiency (32).

Fluorescent pseudomonads reported to decrease take-all can reduce Mn oxides and increase the availability of Mn for plant uptake (99). The addition of Mn cultures of fluorescent pseudomonads increases their antagonism to *G. graminis* (99) and the population of fluorescent Pseudomonads is fours times greater with NH_4-N than NO_3-N (61).

Manganese, as well as other nutrients which influence take-all, appears to exert its greatest influence on the plant's resistance, i.e. slower disease development and restricted lesions (57,101). Resistance to take-all has been correlated with a cultivar's efficiency of Mn uptake; with inefficient cultivars being susceptible and a highly efficient cultivar, resistant (99,100). Efficiency of Mn uptake has been suggested as a criterion for selecting wheat lines with resistance to this disease (33). The interacting effects of Mn and N, especially NH_4-N, in metabolic functions of the plant associated with defense against pathogens are especially applicable for consideration as a functional mechanism of control. Manganese, in contrast to Fe, may have a more direct effect on pathogenesis as an inhibitor of pectolytic (81) and proteolytic (49) exoenzymes required for pathogenesis.

Our attempts to reduce take-all with broadcast applications of $MnSO_4$ or chelated Mn were only marginally effective because of the biological oxidation of the applied Mn to the non-available Mn-oxides. Co-injecting Mn with NH_3 (stabilized with nitrapyrin to inhibit nitrification during seed-bed preparation) provided a convenient means of maintaining Mn availability to reduce disease severity and increase yields (Table 9). Co-injection of other minor elements has not

Table 9. Effect of manganese coinjected with anhydrous ammonia on the yield and take-all of Auburn winter wheat under severe take-all conditions.[a]

Treatment			Grain Yield	Disease Severity
N^b	NI^c	Mn		
----------kg/ha---------				WH^d
0	0	0	480	31
45	0	0	1480	22
45	0.55	0	1824	12
45	0.55	1.0	3264	5

[a] Sandy loam field soil; third consecutive wheat crop.
[b] N applied as NH_3 10 cm deep on 30 cm centers.
[c] NI = nitrapyrin applied with the NH_3 to inhibit nitrification.
[d] WH = % early maturing, white (sterile) heads.

consistently increased yields or reduced take-all in these studies, although a deeper green foliage is observed with Fe and Zn, and yields are increased with Mg under severe Mg deficiency conditions (Table 6).

The most effective means of Mn application and the most effective reduction of take-all we have accomplished has been by Mn seed treatment (0.6 to 1.0 kg/ha Mn) following the preplant application of stabilized NH_3 (Table 10). Stabilized NH_3 suppresses the population of Mn oxidizers in the rhizosphere and the Mn is readily available for uptake by the seedling (50). The effectiveness of Mn application was correlated with its availability for root uptake (32,34,57). Thus, broadcast application has been the poorest; band application, fair; and seed treatment, best. Foliar applications have not been effective in our studies, probably because of poor basipetal translocation in the phloem (3,57,75).

Table 10. Effect of manganese seed treatment on the yield and take-all of wheat fertilized with stabilized anhydrous ammonia.

Treatment		Yield		Take-all	
N[a]	Mn[b]	SL[c]	SiL	SL	SiL
------kg/ha------		----kg/ha----		--WH[d]--	
0	0	800	2010	31	9
0	0.8	940	2010	27	8
65	0	2550	3420	21	2
65	0.8	2810	4020	17	1

[a] N as NH_3 was injected 10 cm deep on 30 cm centers with 0.55 kg/ha nitrapyrin to inhibit nitrification.
[b] $MnCl_2$ was applied to seed in a slurry with 1% methylcellulose.
[c] SL = sandy loam; SiL = silt loam soil.
[d] WH = % early maturing, white (sterile) heads.

Either form of N may aggravate Cu and Zn deficiencies because of their high affinities for N-containing ligands, while only NO_3-N increases the severity of Mn deficiency (4,15,32,70,96). Since NH_4-N fertilizers may have the opposite effect of NO_3-N on Mn (8), Mn appears to have a predominant effect on take-all through NH_4-N metabolism.

CONCLUSIONS

It is apparent that nutrition determines in large measure the susceptibility of wheat and other cereals to take-all. Mineral nutrients are essential metabolic regulators and components for plant growth, but appear to influence disease more indirectly. This may be by off-setting reduced absorption of nutrients, stimulating new root growth, altering host resistance, changing the soil microflora, or reducing the virulence of *G. graminis* (43). Major damage to the root system inflicted by *G. graminis* not only reduces the absorptive capacity of roots, but also restricts the volume of soil encompassed by them. Mineral fertilizers partially offset damage from impaired plant efficiency and facilitate disease escape as new root growth compensates for most damage by disease (41-43). Root growth requires adequate levels of all nutrients but is especially enhanced by P and N.

The marked differences in disease severity observed with different forms of N indicate that the population of *G. graminis* is probably not directly related to disease severity (43,50), although modified virulence by Mn or NH_4-N within the plant or in the rhizosphere is a distinct possibility.

Increased resistance to take-all is most pronounced when plants have a balanced nutrient supply containing NH_4-N The effect of NH_4-N is correlated with increased metabolic activity in roots, modified microbial activity in the rhizosphere, enhanced uptake of specific minor elements, and greater structural and metabolic defenses in the plant (32,42-44,50).

Integration of this information for functional disease control in the crop management program can be readily accomplished by using a balanced nutrition for optimum plant growth, stabilizing N in the NH_4 form by inhibiting nitrification, and enhancing Mn availability by suppressing rhizosphere Mn oxidizers and supplementing base levels of soil Mn. Other cultural practices which reduce take-all, such as the selection of a tolerant variety, delayed seeding, certain crop sequences, and minimized stress appear to influence take-all primarily through their effect on mineral nutrition.

LITERATURE CITED

1. Bolton, J., and Slope, D.B. 1971. Effects of magnesium on cereals, potatoes, and leys grown on the "continuous" cereals site at Woburn. J. Agric. Sci., Cambridge 77:253-259.
2. Bussler, W. 1981. Pages 213-234 in: Copper in soils and plants. J.F. Loneragan, A.D. Robson and R.D. Graham, eds. Academic Press, Sydney.
3. Butler, F.C. 1961. Root and foot rot diseases of wheat. Sci. Bull. No. 77. N.S.W. Dept. Agric., Australia. 98 pp.
4. Chaudhry, F.M., and Loneragan, J.F. 1970. Effects of nitrogen, copper, and zinc fertilizers on the copper and zinc nutrition of wheat plants. Australian J. Agric. Res. 21:865-879.
5. Christensen, N.W., Taylor, R.G., Jackson, T.L., and Mitchell, B.L. 1981. Chloride effects on water potentials and yield of winter wheat infected with take-all root rot. Agron J. 73:1053-1058.
6. Christensen, N.W., Jackson, T.L., and Powelson, R.L. 1982. Suppression of take-all root rot and stripe rust diseases of wheat with chloride fertilizers. Plant Nutrition 1982. Pages 111-116. in: Proc. 9th International Plant and Nutr. Coll., Warwick Univ., England. Commonwealth Agric. Buressix.

7. Christensen, N.W., and Brett, M. 1985. Chloride and liming effects on soil form and take-all of wheat. Agron. J. 77:157-163.

8. Clark, F.E. 1942. Experiments toward the control of take-all disease of wheat and the *Phymatotrichum* root rot of cotton. Tech. Bull. U.S. Dept. Agric. No. 835.

9. Conner, S.D. 1932. Factors affecting manganese availability in soils. Agron. J. 24:726-733.

10. Cook, R.J. 1981. The effect of soil reaction and physical conditions. Pages 343-352 in: The Biology and Control of Take-all. M.J.C. Asher and P.J. Shipton eds. Academic Press, New York.

11. Cook, R.J., Huber, D.M., Powelson, R.L., and Bruehl, G.W. 1968. Occurrence of take-all in wheat in the Pacific Northwest. Pl. Dis. Rep. 52:716-718.

12. Darbyshire, J.F., Davidson, M.S., Scott, N.M., and Shipton, P.J. 1977. Some microbial and chemical changes in soil near the roots of spring barley, *Hordeum vulgare* L., infected with take-all disease. Ecological Bull. (Stockholm) 25:374-380.

13. Darbyshire, J.F., Davidson, M.S., Scott, N.M., Sparling, G.P., and Shipton, P.J. 1979. Ammonium and nitrate in the rhizosphere of spring barley, *Hordeum vulgare* L., and take-all disease. Soil Biol. & Biochem. 11:453-458.

14. Davidson, R.M., Jr., and Goss, R.L. 1972. Effects of P, S, N, lime, chlordane, and fungicides on *Ophiobolus* patch disease of turf. Pl. Dis. Rep. 56:565-567.

15. DeKock, P.C., Cheshire, M.V., and Hall, A. 1971. Comparison of the effect of phosphorus and nitrogen on copper-deficient and -sufficient oats. J. Sci. Fd. Agric. 22:437-440.

16. Dombrovski, N. 1909. Fungi as a cause of the lodging of cereal crops. Khozyoistoo 1909:334-335.

17. Fellows, H. 1929. Studies of certain soil phases of the wheat take-all problem. Phytopathology 19:103.

18. Garrett, S.D. 1938. Soil conditions and the take-all disease of wheat. III. Decomposition of the resting mycelium of *Ophiobolus graminis* in infected wheat stubble buried in the soil. Ann. Appl. Biol. 25:742-766.

19. Garrett, S.D. 1940. Soil conditions and the take-all disease of wheat. V. Further experiments on the survival of *Ophiobolus graminis* in infected wheat stubble buried in the soil. Ann. Appl. Biol. 27:199-204.

20. Garrett, S.D. 1941. Soil conditions and the take-all disease of wheat. VI. The effect of plant nutrition upon disease resistance. Ann. Appl. Biol. 28:14-18.

21. Garrett, S.D. 1944. Root disease fungi. Chronica Botanica Co., Waltham, MA. 177 pp.

22. Garrett, S.D. 1948. Soil conditions and the take-all disease of wheat. IX. Interaction between host plant nutrition, disease escape, and disease resistance. Ann. Appl. Biol. 35:14-17.

23. Garrett, S.D., and Mann, H.H. 1948., Soil conditions and the take-all disease of wheat. X. Control of the disease under continuous cultivation of a spring-sown cereal. Ann. Appl. Biol. 35:435-442.

24. Glynne, M.D. 1950. Close cereal cropping. Effect of cultural treatments of wheat on eyespot, take-all and weeds. Agriculture, London 56:510-514.

25. Glynne, M.D. 1951. Effects of cultural treatments on wheat and on the incidence of eyespot, lodging, take-all and weeds. Ann. Appl. Biol. 38:665-668.

26. Glynne, M.D. 1953. Wheat yield and soil-borne diseases. Ann. Appl. Biol. 40:221-224.

27. Glynne, M.D. 1957. Eyespot and take-all of wheat and barley. Agric. Rev. London 2:10-15.

28. Glynne, M.D. 1965. Crop sequence in relation to soil-borne pathogens. Pages 423-435 in: Ecology of soil-borne pathogens. K.F. Baker and W.C. Snyder, eds. Univ. of California Press, Berkeley, Ca.

29. Glynne, M.D., and Moore, F.J. 1949. Effect of previous crops on the incidence of eyespot on winter wheat. Ann. Appl. Biol. 36:341-351.

30. Goss, R.L., and Gould, C.J. 1967. Some interrelationships between fertility levels and Ophiobolus patch disease in turfgrass. Agron. J. 59:149-151.

31. Graham, J.H., and Menge, J.A. 1982. Influence of vesicular-arbuscular mycorrhizae and soil phosphorus on take-all disease of wheat. Phytopathology 72:95-98.

32. Graham, R.D. 1983. Effects of nutrient stress on susceptibility of plants to disease with particular reference to the trace elements. Adv. in Botan. Res. 10:221-276.

33. Graham, R.D. 1984. Breeding for nutritional characteristics in cereals. Adv. Plant Nutr. 1:57-102.

34. Graham, R.D., and Rovira, A.D. 1984. A role for manganese in the resistance of wheat plants to take-all. Plant Soil 78:441-444.

35. Hageman, R.H. 1980. Effect of form of nitrogen on plant growth. Pages 47-62 in: Nitrification inhibitors - potentials

and limitations. Special Pub. No. 38. American Soc. Agron., Madison, WI. 129 pp.

36. Halsey, M., and Powelson, R. 1981. The influence of NH_4Cl with KCl on antagonism between fluorescent pseudomonads (fp) and *Gaeumannomyces graminis* var. *tritici* (GGT) on the wheat rhizoplane. Phytopathology 71:105.

37. Halsey, M.E. and R.L. Powelson. 1981. The influence of edaphic factors on the suppression of *Gaeumannomyces* root rot of wheat by NH_4Cl + KCl. Phytopathology 71:223.

38. Hornby, D., and Goring, C.A.I. 1972. Effects of ammonium and nitrate nutrition on take-all disease of wheat in pots. Ann. Appl. Biol. 70:225-231.

39. Huber, D.M. 1972. Spring versus fall nitrogen fertilization and take-all of spring wheat. Phytopathology 62:434-436.

40. Huber, D.M. 1976. Plant Disease. Pages 325-327 in: Yearbook of Science and Technology. D.N. Lapedes, ed. McGraw-Hill Co., New York.

41. Huber, D.M. 1978. Disturbed mineral nutrition. Pages 163-181 in: Plant Pathology - An Advanced Treatise. Vol. III. J.G. Horsfall and E.B. Cowling, eds. Academic Press, New York.

42. Huber, D.M. 1980. Role of nutrients in defense. Pages 381-406 in: Plant Pathology - An Advanced Treatise. Vol. V. J.G. Horsfall and E.B. Cowling, eds. Academic Press, New York.

43. Huber, D.M. 1981. The role of nutrients and chemicals. Pages 317-341 in: The Biology and Control of Take-all. P.J. Shipton and M. Asher, eds. Academic Press, London.

44. Huber, D.M. 1981. The use of fertilizers and organic amendments in the control of plant disease. Pages 357-394 in: CRC Handbook of Pest Management in Agriculture. D. Pimental, ed. CRC Press, Inc., Palm Beach, FL.

45. Huber, D.M. 1985. Nutrition and fertilizers. Proc. 1st Int. Wksp. on take-all of cereals. Pages. 327-333 in: Ecology and Management of Soilborne Plant Pathogens. J.F. Kollmorgen (ed.) American Phytopathological Society, St. Paul, MN.

46. Huber, D.M. 1987. Immobilization of Mn predisposes wheat to take-all. Phytopathology 77:1715.

47. Huber, D.M., and Arny, D.C. 1985. Interactions of potassium with plant disease. Pages 467-488 in: Potassium in Agriculture. R.D. Munson, ed. American Society of Agronomy. Madison, WI.

48. Huber, D.M., and Dorich, R.A. 1988. Effect of nitrogen fertility on the take-all disease of wheat. Down to Earth 44:12-17.

49. Huber, D.M., and Keeler, R.R. 1977. Alteration of wheat peptidase activity after infection with powdery mildew. Proc. Am. Phytopathol. Soc. 4:163.

50. Huber, D.M., and Mburu, D.N. 1983. The relationship of rhizosphere bacteria to disease tolerance, the form of N, and amelioration of take-all with manganese. Proc. 4th Inter. Cong. Plant Pathol., Melbourne, Australia. American Phytopathological Society, St. Paul, MN.

51. Huber, D.M., El-Nasshar, H., Moore, L.W., Mathre, D.E., and Wagner, J.E. 1986. Interactions of a peat carrier and potential biological control agents. Phytopathology 76:1104-1105.

52. Huber, D.M., Painter, C.G., McKay, H.C., and Peterson, D.L. 1968. Effect of nitrogen fertilization on take-all of winter wheat. Phytopathology 58:1470-1472.

53. Huber, D.M., and Watson, R.D. 1974. Nitrogen form and plant disease. Ann. Rev. Phytopathol. 12:139-165.

54. Huber, D.M., Watson, R.D., and Steiner, G.W. 1965. Crop residues, nitrogen, and plant disease. Soil Sci. 100:302-308.

55. Huber, D.M., Warren, H.L., Nelson, D.W., and Tsai, C.Y. 1977. Nitrification inhibitors, new tools for food production. BioScience 27:523-529.

56. Huber, D.M., Warren, H.L., Nelson, D.W., Tsai, C.Y., and Shaner, G.E. 1980. Response of winter wheat to inhibiting nitrification of fall-applied nitrogen. Agron. J. 72:632-638.

57. Huber, D.M., and Wilhelm, N.S. 1988. The role of manganese in resistance to plant diseases. Pages 155-173 in: Manganese in Soils and Plants: an International Symposium. R.D. Graham, ed. Waite Agricultural Research Institute, Glen Osmond, Australia.

58. Jouan, B., and Lemaire, J.M. 1974. Modifications in soil biological communities. I. Preliminary study of the effect of incorporating nutritive substrates into the soil and the effect on the evolution of soils phytopathogenic agents. Ann. Phytopathol. 6:297-308.

59. Leeper, G.W. 1970. Six trace elements in soils. Melbourne Univ. Press, Australia.

60. Louw, H.A. 1957. The effect of various crop rotations on the incidence of take-all (*Ophiobolus graminis* Sacc.) in wheat. Sci. Bull. Dept. Agric. Un. S. Africa No. 379. 12 pp.

61. Lucas, P., and Collet, J.M. 1988. Influence de la fertilisation azotee sur la receptivite d'un sol au pietin-echaudage, le developpement de la maladie au champ et les populations de *Pseudomonas* fluorescents. EPPO Bull. 18:103-109.

62.   Lucas, P., and Nignon, M. 1987. Importance of soil type and physicochemical characteristics on relations between a wheat variety (*Triticum aestivum* L. cv Rescler) and two, virulent or hypovirulent, strains of *Gaeumannomyces graminis* (Sacc.) Von Arx and Olivier var. *tritici* Walker. Plant Soil 97:105-117.

63.   MacNish, G.C. 1980. Management of cereals for control of take-all. J. Agric. West Australia 21:48-51.

64.   MacNish, G.C., and Speijers, J. 1982. The use of ammonium fertilizers to reduce the severity of take-all (*Gaeumannomyces graminis* var. *tritici*) on wheat in Western Australia. Ann. Appl. Biol. 100:83-90.

65.   Mattingly, G.E.G., and Slope, D.B. 1977. Phosphate fertilizer and take-all disease of wheat and barley. J. Sci. Fd. Agric. 28:658-659.

66.   McKinney, H.H., and Davis, R.J. 1925. Influence of soil temperature and moisture on infection of young wheat plants by *Ophiobolus graminis*. J. Agri. Res. 31:827-840.

67.   Neales, T. F., and Incoll, L. C. 1968. The control of leaf photosynthesis rate and the level of assimilate concentration in the leaf: a review of the hypothesis. Bot. Rev. 34:107-125.

68.   Nilsson, H.E. 1969. Studies of root and foot rot diseases of cereals and grasses. Ann. Agric. Coll. Sweden 35:275-807.

69.   Olsen, R.A. 1971. Triterpeneglycosides as inhibitors of fungal growth and metabolism. 3. Effect on uptake of potassium, magnesium and inorganic phosphate. Physiologia Pl. 25:503-508.

70.   Ozanne, P.G. 1955. The effect of nitrogen on zinc deficiency in subterranean clover. Australian J. Biol. Sci. 8:47-55.

71.   Ponchet, J., and Coppenet, M. 1962. Influence de la fumare minerale sur le development du pietin-echaudage, *Linocarpon cariceti* B. et Br. Annls. Epiphyt. 13:277-283.

72.   Powelson, R.L., and Jackson, T.L. 1978. Suppression of take-all (*Gaeumannomyces graminis*) root rot of wheat with fall applied chloride fertilizers. Proc. 29th Ann. Fert. Conf. Pacific N.W., Beaverton, OR, pp. 175-182.

73.   Raiz, A.K., and Jarrell, W.M. 1988. Salt-induced solubilization in California soils. J. Soil Sci. Soc. Am. 52:1606-1611.

74.   Reis, E.M., Cook, R.J., and McNeal, B.L. 1982. Effect of mineral nutrition in take-all of wheat. Phytopathology 72:224-229.

75.   Reis, E.M., Cook, R.J., and McNeal, B.L. 1983. Elevated pH and associated reduced trace-nutrient availability as factors

contributing to take-all of wheat upon soil liming. Phytopathology 73:411-413.

76. Roseberg, R.J., Christensen, N.W., and Jackson, T.L. 1986. Chloride, soil solution osmotic potential, and soil pH effects on nitrification. Soil Sci. Soc. Am. J.

77. Rosen, H.R., and Elliott, J.A. 1923. Pathogenicity of *Ophiobolus cariceti* in its relationship to weakened plants. J. Agric. Res. 25:351-358.

78. Roseman, R.S., Phillips, J.D., and Huber, D.M. 1988. Microelement immobilization predisposes wheat to take-all. Phytopathology 78:1504.

79. Rovira, A.D., Graham, R.D., and Archer, J.S. 1985. Reduction in infection of wheat roots by *Gaeumannomyces graminis* var. *tritici* with application of manganese to soil. Pages 212-214 in: Ecology and Management of Soilborne Plant Pathogens. C.A. Parker, A.D. Rovira, K.J. Moore, and P.T.W. Wong, eds. American Phytopathological Society, St. Paul, MN.

80. Roy, K.W., Abney, T.S., and Huber, D.M. 1976. Isolation of *Gaeumannomyces graminis* var. *graminis* from soybeans in the Midwest. Proc. Am. Phytopathol. Soc. 3:284.

81. Sadasivan, T.S. 1965. Effect of mineral nutrients on soil microorganisms and plant disease. Pages 460-469 in: Ecology of Soil-borne Plant Pathogens. K.F. Baker and W.C. Snyder, eds. Univ. of California Press, Berkeley, Ca.

82. Sallans, B.J. 1965. Root rots of cereals. 3. Bot. Rev. 31:505-536.

83. Salt, G.A. 1957. Effects of nitrogen applied at different dates and of other cultural treatments on eyespot, take-all and yield of winter wheat. Field experiment 173. J. Agric. Sci. Cambridge 48:326-335.

84. Sewell, M.C., and Call, L.E. 1925. Tillage investigations relating to wheat productions. Bull. Kansas Agric. Exp. Stn. No. 18. 55 pp.

85. Shipton, P.J. 1972. Take-all in spring-sown cereals under continuous cultivation: Disease progress and decline in relation to crop succession and nitrogen. Ann. Appl. Biol. 71:33-46.

86. Shipton, P.J. 1972. Influence of stubble treatment and autumn application of nitrogen to stubbles on the subsequent incidence of take-all and eyespot. Pl. Pathol. 21:147-155.

87. Shipton, P.J. 1975. Yield trends during take-all decline in spring barley and wheat grown continuously. EPPO Bull. 5:363-374.

88. Smiley, R.W. 1974. Take-all of wheat as influenced by organic amendments and nitrogen fertilizers. Phytopathology 63:822-

825.

89. Smiley, R.W. 1974. Rhizosphere pH as influenced by plants, soils, and nitrogen fertilizers. Soil Sci. Soc. Am. Proc. 38:795-799.

90. Smiley, R.W., and Cook, R.J. 1973. Relationship between take-all of wheat and rhizosphere pH in soils fertilized with ammonium vs. nitrate nitrogen. Phytopathology 63:882-890.

91. Smith, J.D. 1957. Turf disease notes. 1956. J. Sports Turf Res. Inst. 9:233-234.

92. Stetter, S. 1971. Influence of artificial fertilizers on *Ophiobolus graminis* and *Cercosporella herpotrichoides* in continuous cereal growing. Tidsskr. Pl. Av. 75:274-277.

93. Stumbo, C.R., Gainey, P.L., and Clark, F.E. 1942. Microbial and nutritional factors in the take-all disease of wheat. J. Agric. Res. 64:653-665.

94. Syme, J.R. 1966. Fertilizer and varietal effects on take-all in irrigated wheat. Australian J. Exp. Agric. An. Husb. 6:246-249.

95. Taylor, R.G., Jackson, T.L., Powelson, R.L., and Christensen, N.W. 1983. Chloride, nitrogen form, lime, and planting date effects on take-all root rot of winter wheat. Plant Dis. 67:1116-1120.

96. Tills, A.R., and Alloway, B.J. 1981. The effect of ammonium and nitrate nitrogen sources on copper uptake and amino acid status of cereals. Plant Soil 62:279-280.

97. Tsai, C.Y., Huber, D.M., and Warren, H.L. 1978. Relationship of the kernel sink for N to maize productivity. Crop Sci. 17:399-404.

98. Warren, H.L., Huber, D.M., Tsai, C.Y., and Nelson, D.W. 1980. Effect of nitrapyrin and N fertilizer on yield and mineral composition of corn. Agron. J. 72:729-732.

99. Wilhelm, N.S. 1988. Investigations into *Gaeumannomyces graminis* var *tritici* infection of manganese-deficient wheat. Ph.D. Thesis, Univ. Adelaide, South Australia.

100. Wilhelm, N.S., Rovira, A.R., and Graham, R.D. 1987. Aspects of the suppression of take-all of wheat by manganese. Proc. 6th Australoasian Plant Pathol. Soc. Conf. Australoasian Plant Pathology Society. Adelaide, South Australia, 11-15 May.

101. Wilhelm, N.S., Graham, R.D., and Rovira, A.R. 1988. Application of different sources of manganese sulfate decreases take-all (*Gaeumannomyces graminis* var. *tritici*) of wheat grown in a manganese deficient soil. Australian J. Agric. Res. 39: In Press.

INFLUENCE OF NITROGEN AND CALCIUM COMPOUNDS ON DEVELOPMENT OF DISEASE DUE TO *SCLEROTIUM ROLFSII*

Zamir K. Punja
Campbell Institute for Research and Technology,
Davis, California 95616

The soilborne plant pathogen *Sclerotium rolfsii* Sacc. infects well over 500 plant species (2), comprised mainly of dicotyledonous and several monocotyledonous crops. The distribution of the pathogen generally is limited to hot, humid areas of the tropics and subtropics and parts of the southeastern and southwestern United States (2).

The first reported incidence of *S. rolfsii* was by Rolfs in 1892 on tomato. Severe losses have since been reported on many crops, including sugarbeet (17), peanut (2), tomato (35,40), carrot (25) and golf greens (28,29). Control of *S.rolfsii* has been difficult, due in part to the prolific saprophytic growth rate of the fungus in soil, to its ability to colonize and effectively grow on all forms of organic substrates, and to the production of large numbers of sclerotia which can persist in soil for several years.

Primary infections due to *S. rolfsii* occur close to or at the soil surface, where sclerotial germination and mycelial growth are the greatest (31), and generally take place from germinating sclerotia or mycelial growth from an organic food base or infected root (26). Initial symptoms are wilting of the plant, followed by necrosis and often sudden death due to girdling of the stem or primary root. Control measures which are targeted at preventing primary infections near the soil surface appear to have the greatest potential for success. These may include moldboard or flip plowing to bury surface residues and inoculum (11,13,25), application of fungicides to the soil surface (11,25), application of fertilizers containing N or Ca (25), or combinations of these methods (13,25). There are reports of significant disease reduction using antagonistic soil microorganisms (22) or cultural practices (11,25) aimed at reducing inoculum in the upper 10-12 cm of the soil profile.

In this chapter, the influence of N containing compounds, such as ammonium nitrate (NH_4NO_3), ammonium sulfate (($NH_4)_2SO_4$), ammonium bicarbonate (NH_4HCO_3), and urea (NH_2CONH_2), and of Ca containing compounds, such as calcium hydroxide ($Ca(OH)_2$), calcium nitrate ($Ca(NO_3)_2$) and calcium sulfate ($CaSO_4$), on growth of *S. rolfsii*, development of initial infections, and incidence of disease in the field will be discussed. Since the effectiveness of these compounds is influenced by soil texture, moisture and pH, rate and method of application, the host, and disease pressure, these factors will also be considered. The possible mechanisms by which N and Ca containing compounds may reduce disease due to *S. rolfsii* are discussed.

PATHOGEN BEHAVIOR IN SOIL AND IN THE HOST

Central to the discussion on how N and Ca compounds may reduce disease development due to *S. rolfsii* is an understanding of how the pathogen grows and infects. *S. rolfsii* is prevalent primarily in regions with warm, humid climates. Mycelial growth rate and sclerotial germination are greatest at $25\text{-}30^{\circ}C$, with the optimal at $27\text{-}30^{\circ}C$. Optimal soil moisture regimes for sclerotial germination are from 0 to -3 bars; in drier soils (-3 to -10 bars), germination is reduced, and few sclerotia germinate below -10 bars (31). Most mycelial growth and sclerotial germination occur within the upper 8-10 cm of the soil profile; at greater depths, germination is inhibited (31). Survival of sclerotia also is reduced with increasing depth of burial in soil (31) and in warm (>$20^{\circ}C$), moist soil (5). Sclerotial germination is optimal at low pH, within the range of 2.0 to 4.0; at pH above 7.0, germination *in vitro* is inhibited (27).

Sclerotial formation is abundant when the overall nutrient supply, especially C, is high, but appears to be initiated when N levels are low or the C:N ratio is high (23). Sclerotia exude sugars and amino acids in moist soil, particularly after a period of drying, and these compounds increase the activities of microorganisms in the vicinity of sclerotia (10,36). The microfloral population around sclerotia may influence the germination and survival of sclerotia of *S. rolfsii* (10).

Mycelial growth from germinating sclerotia can occur over distances of 3.0-3.5 cm at the soil surface (26). Senescing lower leaves may provide a suitable bridge to increase the distances and the extent to which mycelial growth occurs. Upon contact with the host surface, structures resembling infection cushions are formed (37). Maceration of host tissue results from production of cell wall degrading enzymes, in particular endo-polygalacturonase, in conjunction with secretion of large amounts of oxalic acid (4,30). The oxalic acid sequesters Ca present as Ca pectate in the middle lamella of the cell walls to form calcium oxalate and pectic acid, thus rendering the pectic fraction more susceptible to enzymatic degradation (4,30,33). The oxalic acid also lowers the pH of the tissue to the optimum for enzyme activity (around pH 4.0-4.5) (4). The combined action of oxalic acid and polygalacturonase cause rapid plasmolysis of cells and loss of tissue integrity, resulting in soft-rotting symptoms and tissue maceration which are characteristic of plants infected with the pathogen (2,25).

INFLUENCE OF NITROGEN APPLICATIONS

Disease Development Under Field Conditions

The earliest observations on the influence of nitrogenous fertilizers on development of *S. rolfsii* were made by Leach and Davey (17) in 1934. They showed from results of field trials in California that heavy post-plant applications of anhydrous ammonia (224 kg/ha N) or $(NH_4)_2SO_4$ (198 kg/ha N) reduced disease on sugarbeet from 30% to 7.6 and 9.2%, respectively, and increased sucrose yields per ha. Additional trials in 1936 of preplant applications of $(NH_4)_2SO_4$ and $NH_4H_2PO_4$ (112 kg/ha N), $CaNO_3$ (56 and 112 kg/ha N) and cyanamid (112 and 224 kg/ha N) showed that $CaNO_3$ at 56 kg/ha N and cyanamid at 112 kg/ha N reduced disease by 40-50% and $CaNO_3$ at 112 kg/ha N reduced disease by 70%.

Numerous studies on the influence of N compounds on disease severity due to *S. rolfsii* on a number of crops have been conducted since the early observations by Leach and Davey (17). The results from these reports are summarized in Table 1. Rates of application have been converted to an equivalent N basis, and the percentage disease reduction that was obtained is expressed relative to the level of disease in the control.

It is apparent from Table 1 that the reduction in disease achieved with similar compounds varies with different studies. The response of the crop, the initial level of disease, soil fertility, and time of application of the compound(s) could have influenced the extent to which disease was reduced. For example, extremely low rates of NH_4-N compounds provided a high level of disease control on golf greens (28). In this system, applications were made seven times over the season, providing a continual exposure to the chemical, and the above-ground mycelial growth phase of the pathogen was exposed to the compound. In contrast, over 100 times the rate of NH_4HCO_3 was required to provide a similar level of control on processing carrots (25), where reduction in germination of sclerotial inoculum within the upper

Table 1. Influence of source and level of N on disease due to *Sclerotium rolfsii*.

Crop	Ref.	Nitrogen Source	Total N applied (Kg/ha)[a]	% less disease than untreated check
Sugarbeet	16	$(NH_4)_2PO_4$	112	17
(*Beta vulgaris*)		$(NH_4)_2SO_4$	112	36
Sugarbeet	17	$Ca(NO_3)_2$	56, 112	41, 68
Sugarbeet	17	Cyanamid	112, 224	48, 75
Pan	7	$(NH_4)_2PO_4$	112, 224	84, 92
(*Piper betle*)		$(NH_4)_2SO_4$	112, 224	84, 92
Pan	7	$NaNO_3$	112, 224	84, 92
Annual	19	NH_4NO_3	168	100
larkspur		$(NH_4)_2SO_4$	168	25
(*Delphinium ajacis*)		Cyanamid	168	35
Apple	15	$(NH_4)_2SO_4$	136	8
(*Malus pumila*)				
Tomato	35	$Ca(NO_3)_2$	19	38
(*Lycopersicon*	*esculentum*)			
Tomato	40	NH_4NO_3	56	0
Tomato	40	$Ca(NO_3)_2$	56	0
Sugarbeet	38	$(NH_4)_2SO_4$	120, 240	46, 72
Sugarbeet	38	$CaNH_4NO_3$	120, 240	58, 86
Sugarbeet	38	Urea	120, 240	45, 79
Golf greens	28	NH_4HCO_3	24	95
(*Poa annua*,				
Agrostis tenius)		$(NH_4)_2SO_4$	36	87
Carrot	25	$Ca(NO_3)_2$	212	77 (deep plow)[b]
(*Daucus carota*)				39 (disk)
Carrot	25	NH_4HCO_3	252	89 (disk plow)
				70 (disk)
Carrot	25	NH_4NO_3	336	31 (deep plow)
				23 (disk)
Carrot	25	$(NH_4)_2SO_4$	336	37 (deep plow)
				17 (disk)
Carrot	25	Urea	336	79 (deep plow)
				51 (disk)

[a] All compounds were applied broadcast, with the exception of McClellan (19), Sitterly (35) and Worley and Morton (40) where side-dress applications were made.

[b] Data represent disease reduction in plots that were deep plow or disked prior to application of the compounds.

6-8 cm of a sandy coarse-textured soil profile was required. The results following applications of materials to deep plowed and disked plots illustrate how the combined effects of reduced inoculum levels (in deep plowed plots) and fertilizer applications provide a greater level of disease control than in plots with higher inoculum levels (disked plots) (Table 1).

While direct comparisons amongst the various reports are difficult to make, the results in Table 1 may be summarized as follows: (i) $((NH_4)_2SO_4)$ at rates of 112-336 kg/ha N may provide a 37-87% reduction in disease; (ii) $Ca(NO_3)_2$ at rates of 19-212 kg/ha N may provide a 38-77% reduction in disease; (iii) urea at rates of 120-336 kg/ha N may provide a 45-79% reduction in disease; and (iv) NH_4HCO_3 at rates of 24-252 kg/ha may provide a 89-95% reduction in disease.

Pathogen Growth and Survival

Sclerotium rolfsii can utilize numerous inorganic and organic N compounds as the sole N source when these are added to media at levels within the range of biological utilization (21,31). Growth occurs in the presence of extremely high levels of salts such as NH_4NO_3, $(NH_4)_2SO_4$ and $NaNO_3$ in culture (1,6,14). Germination of sclerotia in agar is similarly not inhibited until high levels are used (1,6). Results from *in vitro* studies are difficult to extrapolate to the field, as rates of application and conversion parameters in soil may result in levels which are orders of magnitude different from those tested *in vitro*.

Mycelial growth and sclerotial germination in natural soil are inhibited by NH_4-N compounds, particularly when the soil pH is above 7.0. Leach and Davey (16,17) first demonstrated the toxicity of $(NH_4)_2SO_4$ and $NH_4H_2PO_4$ at high pH (7.8) and attributed the toxic effects on the fungus to the release of free NH_3. The mediation of toxicity of NH_4-N salts such as NH_4NO_3, $(NH_4)_2SO_4$ and NH_4HCO_3 to *S. rolfsii* by substrate pH has been reported (16,27). Several NH_4-N compounds are fungicidal at high pH; at lower pH (6.0), however, predominance of NH^+_4, which is relatively nontoxic to *S. rolfsii* , resulted in lower inhibition of sclerotial germination (27). In soil, release of NH_3 under neutral to alkaline conditions could occur and would be expected to inhibit growth of *S. rolfsii* (27,34). Urea, for example, has been shown to reduce populations of certain soilborne fungi through release of toxic NH_3 upon hydrolysis (8,39). The effect of a 20 mM solution of urea, added to natural soil, on the germination of sclerotia of *S. rolfsii* is shown in Fig. 1A. Similar results have been obtained with 20 mM solutions of NH_4HCO_3 and $(NH_4)_2CO_3$. Compounds containing the bicarbonate/carbonate anions in addition to

79

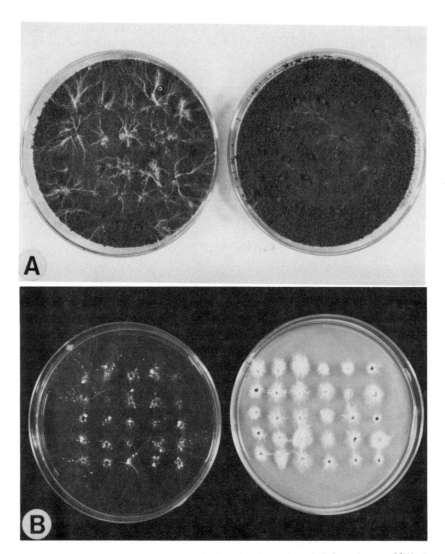

Figure 1. Germination of dried sclerotia of *Sclerotium rolfsii* in the presence of nitrogen or calcium containing compounds. (A) Germination on nonautoclaved fine sandy loam soil, pH 6.8, moistened to approximately -0.5 bar with a 20 mM solution of urea (right) compared with the water control (left). (B) Germination on 1.5% water agar with 50 mM $CaCO_3$ (right) or without (left). Both plates show germination after 72 hr of incubation at $27^{\circ}C$.

NH_3 appear to hold great promise, since these anions are extremely toxic to the pathogen, especially when present in the dissociated form at high pH (27).

Growth of mycelium of *S. rolfsii* on the surface of soil and infection of host tissue can be completely inhibited by urea and NH_4HCO_3 (25). In processing carrots, where root-to-root spread of the pathogen by mycelial growth from infected roots to adjacent healthy roots accounts for rapid increase in the percentage of plants infected (24), inhibition of mycelial growth by applications of N-compounds can provide significant disease control (25).

Leach and Davey (17) reported that application of N compounds did not significantly reduce populations of sclerotia of *S. rolfsii* under field conditions when adjoining treated and untreated plots were compared. However, Punja, et al. (25) showed that one or two applications (each at 84 kg/ha N) of NH_4HCO_3 reduced viability of sclerotia recovered after two months of burial. Enhanced leakage of nutrients from sclerotia exposed to NH_4-N compounds in soil (8) and increased antagonism from soil microorganisms could possibly be one mechanism by which viability of sclerotia was indirectly reduced. Addition of urea (at 40 kg/ha N) in conjunction with *Trichoderma harzianum* reduced viability of sclerotia to a greater extent than either one alone, and also provided a significant level of disease control on peanut (18).

There are a few early reports of increased activities of soil microorganisms in soil following applications of N compounds (2,12). It has been postulated that the inhibitory activities of a number of these soil microbes, particularly bacteria and actinomycetes, to *S. rolfsii* could account, in part, for the observed disease control achieved following applications of N compounds (12,14). Since many organic soil amendments which enhance the levels of microorganisms, including the fraction that are antagonistic to pathogens, have been shown to provide a level of biological control (3), similar results may be expected following applications of the N fertilizers. Field studies in which populations of soil microorganisms before and after N fertilization are monitored have not, however, been conducted to investigate this phenomenon. Long-term laboratory studies also have not been conducted and most investigators have evaluated responses within 3-35 days following fertilizer applications (1,12,14).

Effect on the Host

Virtually all of the reports on the application of N compounds to control *S. rolfsii* have described visible growth enhancement and vigor of the crop, with the presence of darker green leaves and larger roots, when compared to nontreated plots (17,25,28,38). However, more subtle effects on the host, such as anatomical changes in stem or leaf structure, changes in root growth or exudation patterns, have never been described. While healthy, vigorously growing plants appear to be more tolerant to infection by *S. rolfsii* (2), there has been no direct comparison of rates of disease development on crops receiving suboptimal or recommended levels of N with those receiving supplemental doses. Enhanced growth due to N applications was not reported to have any detrimental effects on sugarbeet, carrot, or tomato, especially in soils naturally low in N (17) or in sandy, coarse-textured soils where N may be rapidly depleted by leaching (25). However, on golf greens (28), additional N may be undesirable due to the excessive growth and build-up of thatch, and to increased incidence of other root-infecting pathogens. In this instance, applications of N would not be practical for control of *S. rolfsii* (28).

Recommendations for Disease Control

Applications of N compounds to control disease due to *S. rolfsii* appears to hold promise for a number of host-pathogen systems, in spite of certain inconsistencies. Variability in levels of disease control achieved in response to N applications may be reduced if the following factors are taken into consideration:

i) Applications should be made early in the season, after seedling emergence, to prevent initial infections and secondary mycelial growth of *S. rolfsii* at the soil surface. Materials may be broadcast or side-dressed, and split applications may hold more promise than a single application.

ii) Soil fertility may influence the degree of control achieved. The response may be expected to be more dramatic where soil fertility is low and in the sandier, coarse-textured soils that are common in the southeastern regions of the U.S. (25).

iii) Levels of pathogen infestation and final level of disease incidence can influence the degree of control achieved. With high levels of inoculum, the influence of the added fertilizer may be negligible. Application of N compounds in conjunction with tillage practices that reduce inoculum levels, such as deep plowing, provide better control.

iv) Ammonium and NO_3-N fertilizers appear to provide comparable levels of disease control, with higher rates giving greater disease reduction provided the added N is not detrimental to crop growth.

The mechanisms by which N applications reduce disease due to *S. rolfsii* are: (i) a direct, toxic effect on the pathogen, particularly if ammonia is released; (ii) an indirect effect of predisposition of sclerotia to antagonistic microorganisms; (iii) a direct effect on increasing populations of soil microflora, which in turn reduce activity of the pathogen; and (iv) an indirect effect on reducing susceptibility of the host.

INFLUENCE OF CALCIUM APPLICATIONS

Disease Development Under Field Conditions

Early studies on the application of hydrated lime to control *S. rolfsii* have provided inconsistent results (9). At extremely high rates, small reductions in disease were reported (9,28); in general, however, the level of disease control achieved would not justify the costs of application of the material.

Applications of $Ca(NO_3)_2$ at various rates in some instances provided a significant reduction in disease development on several crops (Table 2). Rates of application of 26-336 kg/ha Ca gave 17-77% disease reduction. Under high inoculum levels and disease pressure, however, $Ca(NO_3)_2$ was ineffective (25,40).

Results with applications of $CaSO_4$ were variable, depending on the rates of application and the level of disease pressure (Table 2). When used in combination with NH_4HCO_3, $CaSO_4$ at 215 kg/ha Ca applied to deep plowed plots gave a 67% reduction in disease on processing carrots (25).

Pathogen Growth and Survival

The effects of Ca containing compounds on mycelial growth and germination, and viability of sclerotia of *S. rolfsii* have been examined with $Ca(NO_3)_2$ (17,25,27), $Ca(OH)_2$ (9), $CaCl_2$ (27), and $CaCO_3$ (17). In all of these studies, there was no significant effect of the compounds on growth of the pathogen or germination of sclerotia, even at high pH. The effect of a 50 mM concentration of $CaCO_3$ on sclerotial germination of *S. rolfsii* is shown in Fig. 1B. These findings suggest that Ca compounds are relatively nontoxic to *S. rolfsii*. The influence on survival of sclerotia in soil has not been investigated.

The addition of Ca to the growth medium resulted in an observable increase in the production of insoluble crystals of calcium oxalate (33). The formation of these crystals resulted from the

Table 2. Influence of source and level of Ca on disease due to *Scelrotium rolfsii*.

Crop	Ref.	Calcium Source	Total Ca applied (Kg/ha)[a]	% less disease than untreated check
Sugarbeet	9	$Ca(OH)_2$	6,000	23
Sugarbeet	17	$Ca(NO_3)_2$	77,154	41, 68
Tomato	35	$Ca(NO_3)_2$	26	38
Tomato	35	$CaSO_4$	43	0
Tomato	40	$Ca(NO_3)_2$	70	0
Tomato	40	$CaSO_4$	56	0
Golf greens	26	$Ca(NO_3)_2$	26	17
(*Poa annua*,		$Ca(OH)_2$	830	7
Agrostis tenuis)				
Carrot	23	$CaCO_3$	336	40[b](Deep plow) 29 (Disk)
Carrot	23	$Ca(OH)_2$	336	30 (Deep plow) 21 (Disk)
Carrot	23	$Ca(NO_3)_2$	336	77 (Deep plow) 39 (Disk)
Carrot	23	$CaSO_4$	336	42 (Deep plow) 3 (Disk)
Carrot	25	$CaSO_4$	215	21 (Deep plow) 13 (Disk)
Carrot	25	$CaSO_4$ + NH_4HCO_3	215 + 168N	67 (Deep plow) 40 (Disk)

[a] All compounds were applied broadcast, with the exception of Sitterly (35) and Worley and Morton (40) where side-dress applications were made.

[b] Data represent disease reduction in plots that were deep-plowed or disked prior to application of the compounds.

sequestering of oxalic acid produced by the fungus (30,32) by exogenous Ca.

Effect on the Host

The reports of successful disease reduction following applications of Ca containing compounds, particularly $Ca(NO_3)_2$ and $CaSO_4$, and the relative nontoxicity of the compounds to the fungus, suggests that mediation of host susceptibility could be involved. The earliest report on the influence of Ca fertilization on reducing susceptibility of the host to *S. rolfsii* was by Mohr and Watkins (20). They reported that applications of $Ca(NO_3)_2$ to tomato prior to inoculation with *S. rolfsii* reduced the rate of disease development. It was postulated that additional Ca stimulated deposition of a phellogen layer due to increased meristematic growth.

Increased levels of Ca in host tissue, present primarily as Ca-pectate in the middle lamella of cell walls, could render tissue more tolerant to the action of cell wall degrading enzymes (4,25,30). From *in vitro* studies, Bateman and Beer (4) and Punja, et al. (30) showed that the presence of Ca (as $CaCl_2$) reduced the activity of polygalacturonase on Ca-pectate or plant tissue, respectively. The inhibitory effect was partially to completely reversed by adding oxalate (4,30). The presence of higher levels of Ca in carrot tissue, achieved either by vacuum infiltration of the material into tissue (30) or by field applications of $Ca(NO_3)_2$ (25), has been reported to reduce the rate of disease development due to *S. rolfsii*. Levels of Ca in the periderm and parenchyma tissues following applications of $Ca(NO_3)_2$ and $CaSO_4$ (336 kg/ha Ca) were significantly higher than in tissues sampled from plots not receiving supplemental Ca (25). In sandy, coarse-textured soils with low cation-exchange capacity, applications of Ca compounds may prove to be beneficial in enhancing normally low tissue Ca levels, and reducing the rate of disease development or incidence of disease due to *S. rolfsii*.

Recommendations for Disease Control

The application of Ca containing compounds for control of disease due to *S.rolfsii* appears to have some potential; however, the following conditions should be met for successful disease reduction:

i) Applications need to be made early in the season to allow uptake of Ca and accumulation in tissues at significantly higher levels prior to infection by the pathogen.

ii) The response to Ca fertilization would be greater in sandy, coarse-textured soils with low cation-exchange capacity than in fine-textured soils.

iii) The reduction in level of disease would be lower with high levels of inoculum or pathogen infestation.

iv) Calcium nitrate provides significantly better levels of disease control than $CaSO_4$. Applications of $Ca(OH)_2$ have not provided significant reductions in disease, even at high rates.

CONCLUSION

Applications of N and Ca containing compounds to reduce development of disease due to *S.rolfsii* have provided variable results, ranging from good control to no control, depending on the crop. Rates of application which have provided significant disease reduction vary considerably, but in general, the time of application and source of N or Ca have been comparable. Disease pressure or pathogen infestation levels markedly influence the response to the compounds, as do soil fertility levels and soil conditions at the time of application. There are several mechanisms by which N and Ca containing compounds could reduce disease, and a single mechanism cannot satisfactorily account for the disease control that is frequently reported.

LITERATURE CITED

1. Avizohar-Hershenzon, Z., and Shacked, P. 1969. Studies on the mode of action of inorganic nitrogenous amendments on *Sclerotium rolfsii* in soil. Phytopathology 59:288-292.
2. Aycock, R. 1966. Stem rot and other diseases caused by *Sclerotium rolfsii*. North Carolina Agric. Exp. Stn. Tech. Bull. No. 174. 202 pp.
3. Baker, K.F., and Cook, R.J. 1974. Biological control of plant pathogens. Freeman Press, San Francisco. 433 pp.
4. Bateman, D.F., and Beer, S.V. 1965. Simultaneous production and synergistic action of oxalic acid and polygalacturonase during pathogenesis by *Sclerotium rolfsii*. Phytopathology 55:204-211.
5. Beute, M.K., and Rodriguez-Kabana, R. 1981. Effects of soil moisture, temperature, and field environment on survival of *Sclerotium rolfsii* in Alabama and North Carolina. Phytopathology 71:1293-1296.
6. Chaudhuri, S., and Maiti, S. 1978. Inhibitory activity of inorganic nitrogen sources to sclerotia of *Sclerotium rolfsii*. Zeitsch. Pflanz. Pflanzen. 85:10-14.
7. Chowdhury, S. 1946. Effect of manuring on the *Sclerotial* wilt of pan (*Piper betle* L.). Indian J. Agric. Sci. 16:290-293.
8. Chun, D., and Lockwood, J.L. 1985. Reductions of *Pythium ultimum*, *Thielaviopsis basicola*, and *Macrophomina phaseolina* in soil associated with ammonia generated from urea. Plant Dis. 69:154-158.
9. Davey, A.E., and Leach, L.D. 1941. Experiments with fungicides for use against *Sclerotium rolfsii* in soils. Hilgardia 13:523-547.

10. Gilbert, R.G., and Linderman, R.G. 1971. Increased activity of soil microorganisms near sclerotia of *Sclerotium rolfsii* in soil. Can. J. Microbiol. 17:557-562.
11. Gurkin, R.S., and Jenkins, S.F. 1985. Influence of cultural practices, fungicides, and inoculum placement on southern blight and Rhizoctonia crown rot of carrot. Plant Dis. 69:477-481.
12. Henis, Y., and Chet, I. 1968. The effect of nitrogenous amendments on the germinability of sclerotia of *Sclerotium rolfsii* and on their accompanying microflora. Phytopathology 58:209-211.
13. Jenkins, S.F., and Averre, C.W. 1986. Problems and progress in integrated control of southern blight of vegetables. Plant Dis. 70:614-619.
14. Johnson, S.P. 1953. Some factors in the control of the southern blight organism, *Sclerotium rolfsii*. Phytopathology 43:363-368.
15. Lavee, S. 1962. The effect of ammonium sulfate and farmyard manure on young M-II apple rootstocks infected with *Sclerotium rolfsii* Sacc. Israel J. Agric. Res. 12:89-90.
16. Leach, L.D., and Davey, A.E. 1935. Toxicity of low concentrations of ammonia to mycelium and sclerotia of *Sclerotium rolfsii*. Phytopathology 25:957-959.
17. Leach, L.D., and Davey, A.E. 1942. Reducing southern Sclerotium rot of sugar beets with nitrogenous fertilizers. J. Agric. Res. 64:1-18.
18. Maiti, D., and Sen, C. 1985. Integrated biocontrol of *Sclerotium rolfsii* with nitrogenous fertilizers and *Trichoderma harzianum*. Ind. J. Agric. Sci. 55:464-468.
19. McClellan, W.D. 1947. Efficacy of certain soil fumigants and fertilizers against crown rot in annual larkspur cause by *Sclerotium rolfsii*. Phytopathology 37:198-200.
20. Mohr, H.E., and Watkins, G.M. 1959. The nature of resistance to southern blight in tomato and the influence of nutrition on its expression. Proc. Am. Soc. Hort. Sci. 74:484-494.
21. Narasimhan, R. 1969. Physiological studies on the genus *Sclerotium II*. Utilization of inorganic nitrogen sources by *Sclerotium rolfsii* (Sacc.) and *Sclerotium oryzae* (Catt.) under protracted incubation. Proc. Ind. Acad. Sci. 49:42-54
22. Punja, Z.K. 1985. The biology, ecology, and control of *Sclerotium rolfsii*. Annu. Rev. Phytopathol. 23:97-127.
23. Punja, Z.K. 1986. Effect of carbon and nitrogen step-down on sclerotium biomass and cord development in *Sclerotium rolfsii* and *S. delphini*. Trans. Brit. Mycol. Soc. 86:537-544.

24. Punja, Z.K. 1986. Progression of root rot on processing carrots due to *Sclerotium rolfsii* and the relationship of disease incidence to inoculum density. Can. J. Plant Path. 8:297-304.

25. Punja, Z. K., Carter, J.D., Campbell, G.M., and Rossell, E.L. 1986. Effects of calcium and nitrogen fertilizers, fungicides, and tillage practices on incidence of *Sclerotium rolfsii* on processing carrots. Plant Dis. 70:819-824.

26. Punja, Z.K., and Grogan, R.G. 1981. Mycelial growth and infection without a food base by eruptively germinating sclerotia of *Sclerotium rolfsii*. Phytopathology 71:1099-1103.

27. Punja, Z.K., and Grogan, R.G. 1982. Effects of inorganic salts, carbonate-bicarbonate anions, and ammonia, and the modifying influence of pH on sclerotial germination of *Sclerotium rolfsii*. Phytopathology 72:635-639.

28. Punja, Z.K., Grogan, R.G., and Unruh, T. 1982. Chemical control of *Sclerotium rolfsii* on golf greens in Northern California. Plant Dis. 66:108-111.

29. Punja, Z.K., Grogan, R.G., and Unruh, T. 1982. Comparative control of *Sclerotium rolfsii* on golf greens in Northern California with fungicides, inorganic salts, and *Trichoderma* spp. Plant Dis. 66:1125-1128.

30. Punja, Z.K., Huang, J.-S., and Jenkins S.F. 1985. Relationship of mycelial growth and production of oxalic acid and cell wall degrading enzymes to virulence in *Sclerotium rolfsii*. Can. J. Plant Path. 7:109-117.

31. Punja, Z.K., and Jenkins, S.F. 1984. Influence of temperature, moisture, modified gaseous atmosphere and depth in soil on eruptive sclerotial germination of *Sclerotium rolfsii*. Phytopathology 74:749-754.

32. Punja, Z.K., and Jenkins, S.F. 1984. Influence of medium composition on mycelial growth and oxalic acid production in *Sclerotium rolfsii*. Mycologia 76:947-950.

33. Punja, Z.K., and Jenkins, S.F. 1984. Light and scanning electron microscopic observations of calcium oxalate crystals produced during growth of *Sclerotium rolfsii* in culture and in infected tissue. Can. J. Bot. 62:2028-2032.

34. Setua, G.C., and Samaddar, K.R. 1980. Evaluation of role of volatile ammonia in fungistasis of soils. Phytopathol. Z. 98: 310-319.

35. Sitterly, W.R. 1962. Calcium nitrate for field control of tomato southern blight in South Carolina. Plant Dis. Rep. 46:492-494.

36. Smith, A.M. 1972. Nutrient leakage promotes biological control of dried sclerotia of *Sclerotium rolfsii* Sacc. Soil Biol. Biochem. 4:125-129.
37. Smith, V.L., Punja, Z.K., and Jenkins, S.F. 1986. A histological study of infection of host tissue by *Sclerotium rolfsii.* Phytopathology 76:755-759.
38. Thakur, R.P., and Mukhopadhyay, A.N. 1972. Nitrogen fertilizaton of sugarbeet in relation to Sclerotium root rot caused by *Sclerotium rolfsii* Sacc. Indian J. Agric. Sci. 42:614-617.
39. Tsao, P.T., and Oster, J.J. 1981. Relation of ammonia and nitrous acid to suppression of *Phytophthora* in soils amended with nitrogenous organic substrates. Phytopathology 71:53-59.
40. Worley, R.E., and Morton, D.J. 1964. Ineffectiveness of calcium nitrate and other calcium sources in reducing southern blight incidence on Rutgers tomato under epiphytotic conditions. Plant Dis. Rept. 48:63-65.

CONTROL OF CLUBROOT OF CRUCIFERS BY LIMING

Robert N. Campbell and Arthur S. Greathead
University of California, Davis, CA 95616
Cooperative Extension Service, Salinas, CA 93901

The role of *Plasmodiophora brassicae* Wor. in clubroot has been recognized since the pathogen was described by Woronin (33) but the disease was well known long before then. Many of the early reports contained speculations about the cause of the disease, and methods to control it (9,19). Liming of the soil was one of the earliest recommendations for controlling clubroot and will be the focus of this paper following a review of the life history of *P. brassicae* and of factors related to liming.

The general life history of *P. brassicae* was established by Woronin (33) but some details, particularly cytological observations, have been established more recently. In the life cycle proposed by Ingram & Tommerup (18), the haploid resting spores produce zoospores that infect the root hairs of the host and form multinucleate primary plasmodia which in turn cleave into zoosporangia. Secondary zoospores are released from the zoosporangia, fuse, and re-infect the root producing the multinucleate secondary plasmodia which invade the host cortical cells and cause the characteristic clubs. Karyogamy occurs within the secondary plasmodium, followed by meiosis and cleavage of the plasmodium into haploid resting spores. The validity of this life cycle has been supported by other types of observations. Synaptonemal complexes, characteristic of meiosis, occur in the plasmodia just before the resting spores form (14). When the release of secondary zoospores was prevented by reducing soil moisture, clubbing was prevented (11). If the fungus does not become a pathogen until it develops into the secondary plasmodial stage, it may be possible to control the disease by arresting the fungus in its non-pathogenic root hair phase.

The addition of lime to raise pH and restore productivity of soils that have become acidic is a widespread practice, regardless of clubroot. Lime is a general term used interchangeably for several forms of Ca and Mg in agricultural applications. The form of lime affects the amount of Ca supplied per unit weight of lime, the neutralizing value of the lime, and the rate of release of Ca. The most widely distributed, naturally occurring form of lime is $CaCO_3$ found in limestone, chalk, marl, or calcite. Limestone reacts with water and CO_2 to form Ca^{++} and $2\ HCO_3^-$. Other forms of lime, such as calcium oxide (quicklime, burnt lime) and especially CaOH (slaked lime, hydrated lime), are more frequently used for clubroot control (3,9,12, 19,27,32). For simplicity, this presentation will refer only to the Ca form of these agricultural limes, although dolomitic lime (a mixture of Ca and Mg carbonates) would probably give the same results.

The determination of pH is the most common test made with soils (26). Soil pH is important because it provides more information about the condition of soil than any other datum (4). For example, there is a correlation between pH and the percentage of base saturation. As the pH is increased from pH 5 to pH 6, the percentage of base saturation increases in mineral soils by about 5% for each 0.1 pH unit (4). The pH is affected also by the nature of the micelle and the ratio of exchangeable bases which vary widely among soils of diverse origins. The correlation between pH and percentage of base saturation, therefore, is not close when diverse soils are compared.

During the years that the clubroot-pH relationships have been studied, the techniques for measuring pH have changed from colorimetric to electrometric making testing faster and more precise. The methods have been standardized by soil scientists (26) and emphasized for plant pathologists (28). The pH value of a soil-water suspension varies inversely with the concentration of neutral salts. Consequently, the pH value of a soil will be affected directly by the neutral salt content of the soil and indirectly by the soil:water ratio selected to prepare the samples for testing. The effect of neutral salts can be avoided by measuring the pH of soil suspended in 0.01 M $CaCl_2$. The pH in 0.01 M $CaCl_2$ is constant regardless of the liquid:soil ratio and it is commonly about 0.5 pH unit lower than for a water suspension (28). While the $CaCl_2$ method gives reproducible values, it does not give a good representation of the conditions to which roots and soil microbes are exposed. Some soil scientists prefer to take the pH of a water-saturated paste which gives a reasonable approximation of the pH in the soil in field conditions, particularly at saturation. The saturated paste is also the starting point for other analyses. Thus, this method is the standard method used by the Cooperative Extension Service Soils Laboratory of the University of California and also by most of the private testing laboratories in California (G. deBoers, pers. comm.). The modern, flat-surface electrodes are admirably suited to this type of measurement. Unfortunately, clubroot researchers have used different methods and some have not reported the method they used so that the pH levels cited in papers are difficult to compare.

FIELD TRIALS OF LIMING FOR CONTROL OF CLUBROOT

Liming of the soil for control of clubroot has been practiced for more than 200 years. Karling (19) and Colhoun (9) have summarized these reports and readers are referred to these reviews. The authors reached similar conclusions which are best expressed in their own words: ". .such measures (including liming and other treatments) have been used with varying degrees of success . . .and the results. . .have often been contradictory." (19) and "It is . . not surprising that conflicting results were obtained even when the same form of lime was used by different workers" (9). In fact, Woronin (33) did not mention lime when he described *P. brassicae* and discussed methods for controlling it.

More recent research papers would not have changed the views of Karling or Colhoun if they were writing today. Hydrated lime [3.4 metric tonnes/ha (t/ha)] applied 6 wk before planting produced a soil pH of 5.0 to 6.5 (determinative method unknown) and relatively poor control of clubroot in Sri Lanka (32). The combination of lime and pentachloronitrobenzene (PCNB), however, provided effective control and improved the growth of the plants. Application of hydrated lime (4.5 t/ha) gave a pH (in 0.01 M $CaCl_2$) of 6.7 and adequate control of clubroot in western Washington, as did lower rates of lime plus PCNB (3). A later report from the same group evaluated variables, such as lime particle size and thoroughness of incorporation into the soil, and their influence on the effectiveness of the "old but unpredictable control measure of liming" (12). Lime [form unspecified in reference, but known to be limestone (A.S. Greathead, personal comm.)] at much higher rates of 22.4 and 44.8 t/ha (half of the total applied in each of two consecutive years) gave a soil pH (saturated paste method) of 7.2 and excellent control of clubroot in Santa Cruz Co., California (29). Fletcher et al. (13) obtained excellent control with 20 t/ha of limestone applied annually for two or three years in different plots. In this case, the soil pH (water suspension method) was increased to >7.4.

Because liming generally is an erratic or partially effective measure, it cannot be relied on exclusively for disease control unless applied at high rates (2,27). These high rates may not be economical or may induce deficiencies of B, Fe or Mn; thus liming is used as one of several control measures, including resistance (if available), chemical treatment of transplants or soil, rotation, and clean seedbeds.

CLUBROOT CONTROL IN THE SALINAS VALLEY

We became concerned about clubroot control when the disease was found in 1978 in the Salinas Valley of Central California. This district is a major crucifer growing area with >26,000 ha of direct-seeded broccoli, direct-seeded or transplanted cauliflower, and other cruciferous crops grown annually. The prospect was that clubroot would spread rapidly because the climate permitted year-round cropping of crucifers and the mechanized nature of the agriculture entailed frequent transport of tillage and harvesting equipment carrying soil and plant parts from field to field over large areas of the valley.

We chose to test the effectiveness of liming for several reasons. It was effective in earlier trials in a nearby area (29) and a large amount of spent lime that had accumulated nearby at a sugar-beet processing factory assured an abundant, cheap supply. This lime was $CaCO_3$ containing approximately 20% impurities (mainly organic material) and 20% moisture. The rates specified in the trials were expressed on the basis of 100% lime.

A single application of 5-10 t/ha in small scale test plots in a heavily infested area gave virtually complete control of clubroot on broccoli for 2 or 3 years (7). During this period the soil pH (saturated paste method) was 6.5 or above in the limed plots. Furthermore, pre-plant application of lime to plots in commercial fields with scattered foci of infection effectively arrested the disease in the following crop. Consequently, spent lime has been applied to most fields at risk from clubroot in the Salinas Valley since 1981. Applications are made annually, particularly if the pH is <6.8, and usually at 4 t/ha of $CaCO_3$. This amount is equivalent to 3 tons of spent lime per acre. The resulting pH levels range from 6.9 upwards with spectacular control of clubroot.

The clubroot-infested area increased rapidly from 1 ha in 1978 to 80 ha in 1981 when liming was started on a large scale. The disease has not been seen on a large scale since that time except on non-limed fields or on isolated patches with a lower pH than the rest of the field. In both situations, application of lime has eliminated the problem for the following crucifer crop. Even though the pathogen has been disseminated throughout the valley judging from the distribution of the small outbreaks, the disease presently is of little economic consequence because of the effectiveness of liming. Recently, clubroot has appeared in the Santa Maria Valley, a coastal valley south of the Salinas Valley and with similar cropping patterns. Lime from a local sugar-beet processing factory is giving similar good control of clubroot (M. Snyder, pers. comm.).

FACTORS INVOLVED IN LIMING TRIALS

A better understanding of the reasons liming is very successful in the Salinas Valley might provide information that could increase the reliability of liming in other areas. Unfortunately, this understanding may be slow to develop because there is no active research on this subject. While the reasons for the consistency of results with the use of lime in the Salinas Valley are not known, other authors have presented many explanations as to why lime was unsuccessful or gave erratic control in their localities and comparisons may be made on these points.

Soil pH has come to play a dominant role in decisions about the need to lime and in evaluations of clubroot control. For example, liming may be indicated if the soil pH is <6.8, the general guideline used in the Salinas Valley. The desired target soil pH is often cited as 7.2 (9,19,27) but this figure may not be appropriate for all soils. For example, poor control was achieved at the target pH of 7.2 in one study (30) and good control was achieved at pH 6.7 in another (7). Doubtless, part of the discrepancy in the literature may be attributed to the use of different methods to determine the pH of the soil. While it is unlikely that one method will be used as a standard everywhere, any method that is chosen should be compared to one of the widely used standards to permit easier assessment of the results.

Another hypothesis related to the effect of pH on clubroot has attributed variable results to the presence of microsites with widely varying pH values in the soil (12,16). Such microsites were detected when small, 0.5 gm samples rather than the usual 15 gm soil samples, were tested. The authors proposed that the fungus was active and infected roots in microsites with a low pH even though a standard pH test indicated the pH was high enough to control the pathogen. This premise is interesting; however, it assumes that the mode of action of lime is to change the pH and affect the pathogen in the soil. If the site of action of lime is within the host root, it is unlikely that the host cells will closely reflect the pH or available Ca in the microsites in the surrounding soil.

Factors other than pH probably play roles that are not fully known. Several factors related to the procedures used in the

94

application of lime have been discussed by others. These variables include: the form of lime selected for the trials, the need for lime to age in the soil before planting, and the necessity to add lime yearly (9,19,27). Hydrated lime and quicklime are regarded as superior to limestone (9,19,27) because they are more 'active' and change the pH and available Ca faster. The activity of these limes doubtless is important in climates where there is a short time between thawing of the soil and planting date. Although these factors may play a role in the control by liming, this role is probably a minor one in the Salinas Valley based on two observations (7). First, limestone gives a high degree of control. This effectiveness and the economics of the available supply of limestone have eliminated the need to compare forms of lime. Secondly, limestone was effective in test plots that were planted a few days after liming and it remained effective for two to three years. Dobson et al. (12) concluded that finer lime particle sizes and more thorough mixing of lime into the soil improved clubroot control. These observations confirmed that the factors important for the action of liming to control clubroot are no different than those given for the correction of soil acidity at least as early as 1913 (31). In the Salinas Valley, agricultural practices normally conform to these recommendations in that the limestone is finely divided and it is thoroughly incorporated into dry soil by disking.

Often factors shown to affect the *P. brassicae*- host interaction in glasshouse studies have been cited to explain the erratic control achieved in the field (6,9,17,19). These factors include: inoculum density, environmental factors such as temperature, light, soil moisture (usually expressed as rainfall frequency and amount), and physical characteristics of the soil. Colhoun (8) proposed that the effectiveness of lime could be overcome by high soil moisture content (70% moisture holding capacity), optimum temperatures (>23 C), and high inoculum density (between 10^5 and 10^7 or more spores/gm soil). Our plots were irrigated by sprinklers every other day until the stand was established in a soil with up to 10^6 spores/gm soil. We suggest, therefore, that these factors are of minor importance for understanding our results. On the other hand, soil temperatures may play a role. A set of soil temperature measurements was made in 1980 in the Salinas clubroot plots (24). The weekly maximum soil temperatures at 10 cm depth were above 23 C for several weeks but the weekly mean soil temperature was <23 C. Thus, cool soil temperatures produced by the coastal climate of the Salinas Valley may be a factor that permits liming to control clubroot. Despite this, we do not attach great importance to this hypothesis because liming has given erratic control of clubroot in other areas, e.g. western Washington and northern Europe, where the climate is cool but otherwise quite different from here.

Recent studies have shown that additional factors need to be evaluated in field trials. Because it is impossible to change the pH of soil by liming without also changing the Ca level, we agree and believe that changes in Ca may be as important as changes in pH. Dobson et al. (12) tested several N fertilizers applied to limed or non-limed soils. There was less clubroot on non-limed soils fertilized with $Ca(NO_3)_2$ than with other fertilizers. The authors attributed this reduction in clubroot to the differential uptake of NO_3-N which would increase the soil pH although this factor was not measured. Fletcher et al. (13) applied $CaCO_3$, Na_2CO_3, $CaSO_4$, and Na_2CO_3 annually for 3 years. The carbonates increased the pH to 8.0 and controlled clubroot. The Ca sulfate treatment reduced the pH to 6.7-6.8 from pH 7.0-7.1 in the check plots yet there was less clubroot in the $CaSO_4$ treatments than in the check. The authors speculated that some factor other than soil pH influenced the amount of disease. Clubroot could have been reduced by the added Ca in both cases.

Another factor related to soil pH and Ca changes may be the types of soil in the Salinas Valley. These soils, developed in an arid or semi-arid climate, tend to have an adsorbed cation complex dominated by Ca and Mg rather than by Ca and H complex that is found in soils developed in wetter conditions (4). The latter soils have been leached and are more acidic. Liming of soils formed in an arid climate may produce different results than liming of soils formed in wetter conditions as far as the balance of cations, the percentage of base saturation, or the availability of Ca. The liming trials in the Salinas Valley are probably the first to be done in arid-climate soils and the success may be explained on the basis of the interaction between pH and available cations (Ca and Mg) (7). Clubroot was controlled when the extractable cations exceeded 14 meq/100 g of soil and the pH was from 6.7 to 7.2. Below pH 6.7, a much higher extractable cation level may be needed to reduce clubroot. Above pH 7.2, the level of extractable cations may be less important than pH in controlling clubroot. The figure of 14 meq of Ca plus Mg is a first approximation based on tests in one soil type and needs further testing. The determination of extractable cations is an appropriate test for clubroot trials because it measures the cations available to the plant roots. The percentage of base saturation and ratios of exchangeable bases also may be useful criteria to compare soils but they have not been done.

MODE OF ACTION OF LIME

Despite the long record of using lime to control clubroot, there is surprisingly little information about its mode of action. "In no area of clubroot control is so little known of the mode of action of a reasonably effective controlling agent than of the effect of calcium containing materials" (5). The role of Ca and Mg in soil chemistry is complex (11). It is nearly impossible to do experiments in which pH and Ca are manipulated independently of each other in a soil medium. Nutrient culture systems, are used by plant physiologists to study mineral nutrition, overcome these difficulties.

A solution culture technique in which roots were exposed to resting spores in a nutrient solution contained in small vials was used by MacFarlane (21). His phosphate-buffered nutrient solution reduced infection at the standard concentration, but not at 1/5th or 1/25th of that concentration. Infection was reduced at pH 7.0 compared to pH 5.0 or 6.0. No other pH values were tested so the pH limits for infection were not established. He thought pH changes, rather than the associated changes in cations, were responsible for the effects of lime. He also recognized clearly that it was impossible to prove the point with phosphate-buffered solutions because changes in pH changed the free Ca content through the formation of insoluble Ca phosphates.

Palm (25) added phosphate-buffered nutrient solutions to plants growing in sand and tested the effects of Ca, K, and pH on primary infection and on gall development by *P. brassicae*. There was a very narrow, sharp peak in the number of primary infections at 0.75 to 1 .0 mM Ca or at 4 to 7 mM K when the pH was 5.9. High Ca concentrations had less effect on gall size than on primary infections and potassium had no effect on gall size. The effect of pH on primary infection was as sharply defined as the effect of Ca. A peak of infection occurred at pH 4.5 to 5.2 with sharp declines on either side of that range and gall development responded similarly. Palm also studied the effects of other elements but these will not be discussed here. He concluded that if one followed the empirical rule of liming to pH 7.2, control might fail because pH is only one factor and high Ca levels might not be effective in some conditions such as in the presence of B deficiency, he also reasoned that the first effect of liming was to reduce primary infection by direct effect of Ca or by shifting pH beyond the optimum for infection. He clearly recognized that Ca or pH might have different effects upon the pathogen in different stages of its life cycle.

More recently, the effects of Ca and pH were separated by using the Goode series of buffers (15) that were not available to Palm or MacFarlane (23). There was an inverse relationship between pH in the range of 6.2 to 7.2 and the Ca level at which abundant primary infections occurred. The number of primary infections was decreased by Ca in excess of 25 mM at pH 6.2 down to 1.5 mM at pH 7.2. Moreover, the maturation of zoosporangia and release of secondary zoospores was delayed or prevented at the higher levels of Ca at which normal numbers of primary infections occurred. This effect of Ca was more pronounced at pH 7.2 than at lower pH levels. Furthermore, clubbing was reduced or prevented at the Ca levels at which primary infections occurred but did not mature. Thus, both Ca and high pH affected the host pathogen interaction, presumably at one or more stages of development after infection by the primary zoospores. The responses to Ca and pH were quite different from those reported by Palm, presumably because of the different buffer systems that were used. Nevertheless, the conclusion that both Ca and pH have an effect was supported. Dixon & Webster (10) used a similar nutrient culture system and confirmed that both pH and Ca have effects on infection and post-infection development of *P. brassicae*. A role for Ca in the post-infection development of *P. brassicae* is supported also by the demonstration that incorporation into roots is pH-dependent (23). Broccoli seedlings at pH 6.2 incorporated less Ca from solutions with 1.5 or 7.5 mM Ca than seedlings at pH 6.8 or 7.2.

The site of action of Ca has been debated. Some early authors maintained that liming reduced the survival of resting spores (9); however, recent authors have not accepted this view. MacFarlane (20) showed infectivity declined more rapidly in acidic soil than in alkaline soil. Fletcher et al. (13) believed that Ca or high pH did not affect inoculum survival in field soils. Myers and Campbell (23) subjected resting spores to molar Ca or Na at three pH levels for a week and noted a reduction in viability no greater than 10%. Even this reduction could have been caused by residual Ca that interfered with spore germination. Because molar Ca far exceeds the concentration in agricultural soils, it is unlikely that Ca *per se* can affect resting spore viability. In this case, the effects of pH and Ca are more likely to be expressed at one or more stages between spore germination and completion of the life cycle and just as likely to be expressed within the host cells as in the soil surrounding the roots. Furthermore, the relatively small change of pH of the soil, e.g. from pH 6.0 to 7.2, would have more effect on the uptake of Ca by roots than on the intracellular pH which is actively buffered by the host cells. Thus, more attention to the effect of Ca to the post-infection development of the pathogen in the host is warranted (23).

The role of Ca in the host-parasite interrelationship is unknown but Ca plays a number of important roles in the plant cell aside from its involvement in intercellular pectins. Ca very likely is a cellular messenger which affects the regulatory mechanisms in plant cells (22). There are calmodulins (acidic, low molecular weight, Ca-binding proteins) in plants and Ca- or calmodulin-regulated enzymes have been identified. The three enzymes of this type (NAD kinase, ATPase, and protein kinases) may be in the soluble fraction of the cell or associated with membranes. The membrane association is of great interest because of the membrane bound nature of the thalli of *P. brassicae* in host cells (1).

CONCLUSION

Liming the soil for the control of clubroot has given erratic results in most places but it has given excellent disease control in the Salinas Valley. Much research still is needed to elucidate the mode of action of lime, soil pH and other factors in controlling clubroot. This understanding could be important to explain the erratic control achieved in most areas of the world and might lead to the more extensive use of this non-toxic, non-polluting control technique in preference to the application of fungicides or soil fumigants. We hope that this chapter will stimulate new approaches to research and to understanding and improving the control of clubroot in areas where it is a continuing problem.

LITERATURE CITED

1.   Aist, J.R., and Williams, P.H. 1971. The cytology and kinetics of cabbage root hair penetration by *Plasmodiophora brassicae*. Can. J. Bot. 49:2023-2034.
2.   Anon. 1984. Clubroot. Minist. Agric., Fish. & Food. Leaflet 276. 7 pp.
3.   Anderson, W.C., Gabrielson, R.L., Haglund, W.A., and Baker, A.S. 1976. Clubroot control in crucifers with hydrated lime and PCNB. Plant Dis. Rep. 60:561-565.
4.   Brady, N.C. 1974. The nature and properties of soil. 8th ed. MacMillan Publ. Co., N.Y. 639 pp.
5.   Buczacki, S.T. 1977. Chemical and cultural control of *Plasmodiophora brassicae*. pp. 36-43. Proc. Woronin + 100 Conference. Univ. of Wisconsin, Madison, WE. 159 pp.
6.   Buczacki, S.T., Ockendon, J.G., and Freeman, G.H. 1978. An anlysis of some effects of light and soil temperature on clubroot disease. Ann. Appl. Biol. 88:229-238.

7. Campbell, R.N., Greathead, A.S., Myers, D.F., and de Boer, G.J. 1985. Factors related to control of clubroot of crucifers in the Salinas Valley of California. Phytopathology 75:665-670.
8. Colhoun, J. 1953. A study of the epidemiology of club-root disease of cabbage. Ann. Appl. Biol. 40:262-283.
9. Colhoun, J. 1958. Club root disease of crucifers caused by *Plasmodiophora brassicae* Woron. Phytopathological Paper No. 3. Commonwealth Mycol. Instit., Kew. 108 pp.
10. Dixon, G.R., and Webster, A. 1986. Host nutrition in relation to clubroot (*Plasmodiophora brassicae*) control. (Abstr.). Hortscience 21:753.
11. Dobson, R.L., and Gabrielson, R.L. 1983. Role of primary and secondary zoospores of *Plasmodiophora brassicae* in the development of clubroot in Chinese cabbage. Phytopathogy 73:559-561.
12. Dobson, R.L., Gabrielson, R.L., Baker, A.S., and Bennet, L. 1983. Effects of lime particle size and distribution and fertilizer formulation on clubroot disease caused by *Plasmodiophora brassicae*. Plant Dis. 67:50-52.
13. Fletcher, J.T, Hims, M.J., Archer, F.C., and Brown, A. 1982. Effects of adding calcium and sodium salts to field soils on the incidence of clubroot. Ann. Appl. Biol. 100:245-251.
14. Garber, R.C., and Aist, J.R. 1979. The ultrastructure of meiosis in *Plasmodiophora brassicae* (Plasmodiophorales). Can. J. Bot. 57:2509-2518.
15. Gueffroy, D.E. 1975. Buffers. Calbiochem-Behring Corp., San Diego, CA. 24 pp.
16. Haenseler, C.M. 1937. Control of clubroot of crucifers. (Abstr.). Phytopathology 27:130.
17. Hamiloton, H.A., and Crete, R. 1978. Influence of soil moisture, soil pH, and liming sources on the incidence of clubroot, the germination and growth of cabbage produced in mineral and organic soils under controlled conditions. Can. J. Plant Sci. 58:45-53.
18. Ingram, D.S., and Tommerup, I.C. 1972. The life history of *Plasmodiophora brassicae* Woron. Proc. R. Soc. Lond. B. 180:103-112.
19. Karling, J.S. 1968. The Plasmodiophorales. Hafner Publ. Co., New York. 256 pp.
20. MacFarlane, I. 1952. Factors affecting the survival of *Plasmodiophora brassicae* wor. in the soil and its assessment by a host test. Ann. Appl. Biol. 39:239-256.

21. MacFarlane, I. 1948. A solution culture technique for obtaining root-hair, or primary, infection by *Plasmodiophora brassicae*. J. Gen. Microbiol. 18:720-732.

22. Marme, D. 1985. Role of calcium in the regulation of plant metabolism. Pages 1-8 in: Molecular and cellular aspects of calcium in plant development. A.J. Trewavas, ed. Plenum Press, New York. 452 pp.

23. Myers, D.F., and Campbell, R.N. 1985. Lime and the control of clubroot of crucifers: Effects of pH, calcium, magnesium, and their interactions. Phytopathology 75:670-673.

24. Myers, D.F., Campbell, R.N., and Greathead, A.S. 1983. Thermal inactivation of *Plasmodiophora brassicae* Worm. and its attempted control by solarization in the Salinas Valley of California. Crop Prot. 2:325-333.

25. Palm, E.T. 1963. Effect of mineral nutrition on the invasion and response of turnip tissue to *Plasmodiophora brassicae* wor. Contr. Boyce Thompson Inst. 22:91-112.

26. Peech, M. 1965. Hydrogen-ion activity. pp. 914-932 in C.A. Black. ed. Methods of soil analysis. Am. Soc. Agron., Madison, WI.

27. Sherf, A.F., 1976. Clubroot of cabbage, cauliflower, and broccoli. Cornell University Cooperative Extension Leaflet. 2 pp.

28. Smiley, R.W., and Cook, R.J. 1972. Use and abuse of soil pH measurement. Phytopathology 62(2):193-194.

29. Welch, N., Greathead, A.S., Inman, J., and Quick, J. 1976. Club root control in Brussels sprouts using lime for pH adjustment. Ca. Agric. 30(4):10-11.

30. Wellman, F.L. 1930. Clubroot of Crucifers. U.S.D.A. Tech Bull, 181. 31 pp.

31. Whitson, A.R., and Weir, W.W. 1913. Soil Activity and Liming. Wisc. Agric. Exp. Stn. Bull. 230. 33 pp.

32. Wimalajeewa, D.L.S. 1975. Field investigations on the control of club root of cabbage in Sri Lanka. Ann. Appl. Biol. 79:321-327.

33. Woronin, M.S. 1878. *Plasmodiophora brassicae*: The cause of cabbage hernia. Phytopathol. Classic No. 4. American Phytopathological Society, St. Paul, MN. 32 pp.

REDUCING THE SEVERITY OF BACTERIAL SOFT ROT BY INCREASING THE CONCENTRATION OF CALCIUM IN POTATO TUBERS

Arthur Kelman[1], Raymond G. McGuire[2], and Kuo-Ching Tzeng[3]

[1]Department of Plant Pathology
University of Wisconsin
Madison, Madison, WI 53706
[2]Department of Plant Pathology
Cornell University
Ithaca, NY 14853
[3]Department of Plant Pathology
National Chung Hsing University
Taichung, Taiwan.

The bacterial soft rot diseases of potato (*Solanum tuberosum* L.) caused by the "carotovora" group in the genus *Erwinia*, including *E. carotovora* subsp. *atroseptica* (van Hall) Dye (Eca), *E. carotovora* subsp. *carotovora* (Jones) Bergey *et al.* (Ecc) and *E. chrysanthemi* Burkholder, McFadden, and Dimock (Echr), have frequently caused serious losses worldwide both before and after harvest (16,65). At present no effective chemical controls or highly resistant cultivars are available. In general, standard cultural practices or the use of pathogen-free seed have been relatively ineffective in reducing the losses that occur periodically under field conditions and in the post-harvest period (65). This chapter summarizes field and laboratory research completed in Wisconsin in the period from 1979 to 1986 to determine the role of Ca nutrition on severity of bacterial soft rot of potato tubers (58,59,78,79) and reviews research related to our findings.

Numerous studies reporting the influence of mineral nutrition on physiogenic and biotic diseases of plants have been summarized by Huber (43). His extensive review of this topic lists several reports on the association between Ca and changes in resistance of plants to a number of disease agents. In particular, changes in Ca concentrations in plant tissues have been shown to influence localization of virus

102

infections (82), as well as to moderate disease severity in various crops attacked by fungal pathogens (22,23,26,27,34,36,60,75). The application of Ca affects tissue softening in pickles (13) and cell wall integrity in cucumber (*Cucumis sativus* L.) roots (46) and also reduces the severity of a number of abiotic diseases including blossom end rot of tomato (*Lycopensicon esculentum* Mill.) (38), blackheart of celery (*Apium graveolens* L. var *dulce* DC.) (37), and bitter pit of apple (*Malus sylvestris* Mill.) (6). In potatoes, applications of Ca have been shown to reduce development of internal brown spot in tubers and sub-apical necrosis of sprout (19, 33, 41, 78, 79, 80).

An increase in the concentration of Ca in plant tissue may result also in an increase in resistance to bacterial diseases such as bacterial wilt caused by *Pseudomonas solanacearum* Smith (68; Kelman, unpublished data), and bacterial soft rot of Chinese cabbage [*Brassica pekinensis* (Lour.) Rupr.] (35, 64) and bean (*Phaseolus vulgaris* L.) (66). An in-depth study of the influence of Ca on bacterial soft rot of potato tubers had not been reported when our investigation was initiated, however.

Natural soils usually contain concentrations of Ca adequate for normal growth requirements of plants; the concentration of Ca reported to be present in a midwestern sandy loam soil was about 4.5×10^3 kg/ha. In one study of the nutrients removed from an acre of soil producing 6.8×10^3 kg of potato tubers, 19.5 kg of Ca were deposited in stems and foliage; in contrast only 0.45 kg was present in the tubers (32). Furthermore, the small amount of Ca present in tubers was not evenly distributed; the highest concentration was in periderm and cortex tissue (11). Ca in medullary tissue of potato tubers was very low, in the range of 0.01 to 0.08% dry weight (DW).

On the basis of analyses of large numbers of plants of many species, Ca concentrations ranging from 0.2 to 4.0% DW have become accepted as the typical range for concentrations of this macronutrient in leaves and stems of crop plants (6). Although concentrations of Ca in potato tubers are characteristically much lower than in leaves and stems, the point at which tubers may be considered to be deficient has not been determined (77). There is evidence, however, that as the content of Ca declines in tuber medullar tissues from 0.018 to 0.014% DW, the incidence of the internal brown spot deficiency symptom can double (19).

Although intracellular concentrations of K may range between 20 and 100×10^3 M, concentrations of Ca within the cell must be maintained between 10^{-5} and 10^{-8} M to prevent interference with cellular functions. Because calcium is actively excluded from the living cells of the phloem, very little Ca moves in this tissue (18,28). Calcium, therefore, moves throughout the plant primarily in the xylem and preferentially toward meristematic and transpiring tissue. Once deposited in a given tissue there is almost no redistribution. Internal leaves, fruits or storage organs that do not transpire are likely to suffer from variations in the supply of Ca resulting in localized deficiencies within the plant (6,17,45). These deficiencies may occur in fruits and storage organs even when there is an adequate supply of Ca in the soil solution and plants show no apparent symptoms of Ca deficiency in their foliage. In addition, these deficiencies can occur when high concentrations of other cations such as Mg compete with Ca for uptake by the root system.

FACTORS INFLUENCING SUSCEPTIBILITY OF POTATO TUBERS TO BACTERIAL SOFT ROT

Degree of esterification and solubility of cell wall pectin can affect the susceptibility of tubers to bacterial soft rot (81). Pagel and Heitefuss (63) reported that as the amount of anhydrous galacturonic acid in cell walls and cell wall binding capacity for Ca increased, the susceptibility of tubers to bacterial soft rot decreased. In Russet Burbank and Superior tubers, the amount of galacturonic acid recovered from cell walls was highest in tubers with the highest content of Ca (59). Whether this increase is attributable to increased synthesis of this cell wall component or to a conversion from soluble pectins that would otherwise be lost in preparation of cell wall material has not been determined. Rainfall and unfavorable storage conditions could increase the amount of soluble pectin in the cell walls and facilitate tissue maceration by soft rot bacteria. In addition, dry matter (1,9,40,71,78), phytoalexin formation (54), polyphenol content (1,71,76,78) and polyphenol oxidase activity (52), water potential (2,44), blackspot susceptibility (84), membrane permeability (83), sucrose content (83) and reducing sugar content (62) also have been reported to influence or to be related to susceptibility of potato tubers to bacterial soft rot. However, Cother and Cullis (25) have obtained data indicating that neither high content of reducing sugars nor high water potential enhances susceptibility of tubers to bacterial soft rot. Additional studies are needed to determine how differences in cultivars and storage period may be affecting the results obtained by different researchers in their studies on the relationship of these and other factors to the severity of bacterial soft rot in tubers.

INCREASED RESISTANCE TO BACTERIAL SOFT ROT IN POTATO
TUBERS WITH AN INCREASED CALCIUM CONTENT

In studying the influence of tuber Ca content on development of bacterial soft rot, tubers inoculated with Eca were usually incubated for 4 days in a mist chamber at a temperature of 20 C (12,53). It should be emphasized that these conditions were more favorable for soft rot development than those encountered during commercial storage or in transit (12,53). A constant mist produces a continuous film of water over surfaces of the tubers; under such conditions, the tubers become anaerobic within 3 hr (15) and disease severity is greatly enhanced. In storage, low oxygen conditions often develop when tubers are covered with a film of water (15) as the result of condensation, or in transit when freshly washed potatoes are not dried before shipment.

Infiltration of Tubers with Calcium Solutions

Initial studies to determine the influence of Ca on susceptibility of potato tubers to bacterial soft rot involved experiments in which Ca salts were infiltrated into tubers by a vacuum procedure (58). Suspensions of Eca were added to the various Ca solutions at the time tubers were infiltrated. Vacuum-infiltration of Ca solutions into potato tubers increased the Ca content in peel and medullar tissues significantly. Chloride, NO_3, SO_4 and gluconate salts of Ca were tested, but the low solubilities of $CaSO_4$ and Ca-gluconate prevented the use of these two salts except at low concentrations.

Increasing the concentration of Ca within the tubers decreased severity of decay after 60 hr in the mist chamber, although all tubers had been uniformly inoculated with Eca (Fig. 1). Percent surface area decayed (SAD) was reduced from 93.3 to 15.0% over the range of concentrations of Ca infiltrated (up to 6000 mg/L). At the high Ca levels, these low percent SAD readings were similar to those observed in control tubers that were not inoculated, but simply infiltrated with water. At a Ca concentration of 12,000 mg/L, no decay was observed; however, at this high salt concentration, viability of bacterial cells in suspension was reduced and even the low background level of decay observed in the non-inoculated tubers was eliminated. When the soft rot data and the tuber Ca levels were analyzed statistically, the negative correlation between percent SAD and tuber Ca content was evident (r= -0.981, P < 0.05).

To determine whether reduction in soft rot severity was attributable to a specific action of Ca or whether it was merely a more general salt effect, tubers were vacuum-infiltrated with Eca suspended

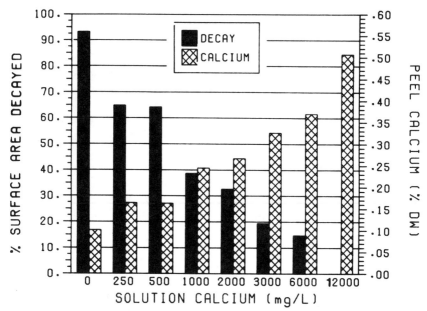

Figure 1. Percent surface area decayed in relation to peel calcium content of Superior tubers vacuum-infiltrated with *Erwinia carotovora* subsp. *atroseptica* in calcium nitrate solutions. Tubers were immersed in a suspension containing $2x10^6$ cfu/ml for 2 hr, the first hour of which was at 100 mm Hg air pressure, followed by incubation in a constant mist chamber at 20 C for 60 hr. Calcium was measured by inductively coupled plasma-optical emission spectrometry.

in solutions of Na, K, Sr, Mg, or $Ca(NO_3)_2$. The monovalent cations, Na^+ and K^+, were least effective in reducing decay when infiltrated into tubers, whereas the divalent cations, Sr^{++}, Mg^{++}, and Ca^{++}, produced a significantly greater reduction in decay. Of these cations, treatment with Ca was the most effective.

Increasing the concentration of Ca within the potato tuber artificially by vacuum-infiltration resulted in a significant decrease in percent SAD. Vacuum-infiltration proved to be a very effective method of increasing tuber Ca concentrations. Non-inoculated tubers subjected to this treatment did not appear to be damaged and, when surface-sterilized with solutions of NaOCl prior to treatment, remained free of decay up to one week, even under mist chamber conditions. Only at the very high concentration of 12,000 mg/L of Ca was the population of Eca adversely affected in this procedure. Thus, it was concluded that the decrease in severity of soft rot associated with vacuum-infiltration into potato tubers of Ca solutions was attributable primarily to Ca effects on tuber tissue resistance to bacterial maceration (58,59).

Field Trials Involving Soil Applications of Calcium

Field trials (in cooperation with Drs. Paul Fixen and Keith Kelling of the Department of Soil Science, UW-Madison) were designed to evaluate the influence of different sources of Ca and methods of placement on the content of Ca in tubers and on yield and quality (72,73). The sandy loam soil on which these trials were completed had a low cation exchange capacity and less than 1000 kg of Ca per hectare. A range of tuber Ca concentrations was obtained through applications of $MgSO_4$ to antagonize Ca uptake of $Ca(NO_3)_2$ and $CaSO_4$ to augment Ca already present in the soil. Tubers from these field plots were tested in the mist chamber assay procedure for susceptibility to soft rot (12,53).

The fertilizer treatments applied the first year were effective in producing subplots with a wide range of soil Ca and Mg. The ratio of soil Ca to Mg was increased nearly eightfold across this range. The increased availability of soil Ca resulted in an increase in uptake of the cation by the Russet Burbank potato plants. The greatest amount of Ca was deposited in the foliage 0.190 to 1.188% DW, and in no treatments were the leaves deficient in this mineral. Tuber Ca also increased as soil Ca increased, both in the peel and in the medullar tissues. Peel Ca ranged from 0.057 to 0.277% DW and medullar Ca from 0.011 to 0.062% DW across all eighteen treatments (58).

The increase of tuber Ca was correlated with a decrease in the percent SAD. Percent SAD was reduced from 43.5% in tubers from Mg subplots receiving no supplemental Ca to 19.4% in tubers from a subplot receiving the greatest level of Ca fertilization. The correlations between percent SAD and peel Cag and percent SAD and medullar Ca were r = -0.933 and -0.948, respectively (P< 0.05).

The following year results were similar to those obtained previously for both tuber Ca and tuber susceptibility to bacterial soft rot (Fig. 2). Percent SAD was again reduced this time from 82.5% in the most susceptible tubers (lacking any Ca fertilization) to 48.5% in tubers from the subplot that received the highest amount of Ca. Surface decay was inversely correlated with concentrations of Ca in peel and medullar tissue. This second year, the correlation between percent SAD and peel Ca was r = -0.917, and between percent SAD and medullar Ca was r = -0.880 (P<0.05). Thus, initial field studies confirmed the earlier observations of an inverse correlation between the concentration of Ca and the severity of bacterial soft rot in tubers (58).

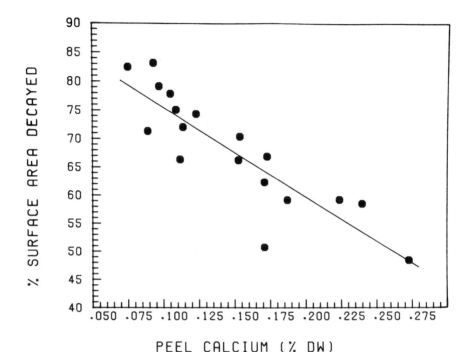

Figure 2. Percent of potato tuber surface area decayed in relation to the percentage of calcium in peel of Russet Burbank tubers following inoculation with *Erwinia carotovora* subsp. *atroseptica.* Tubers were immersed for 20 min. in an inoculum suspension containing 10^6 cfu/ml, followed by incubation in a mist chamber at 20 C for 96 hr. Calcium was measured by inductively coupled plasma-optical emission spectrometry.

Influence of Different Calcium Sources on Soft Rot Susceptibility

The effects of various Ca sources on peel Ca concentration and bacterial soft rot susceptibility of potato tubers were evaluated over three years (72,73). Five Ca sources were applied the first two years by a preplant strip application method at a rate of 252 kg/ha of Ca. In both years, concentrations of tuber Ca was increased by treatments with the different Ca sources. Tubers from plots that received applications of Ca also were consistently less susceptible to bacterial soft rot than tubers from control plots. Of the fertilization treatments, plants receiving applications of sieved or granular $CaSO_4$ produced tubers relatively higher in Ca than did those plants which received lime, and tubers of the sieved $CaSO_4$ treatment were the least susceptible to decay.

A number of different Ca sources were evaluated again in the third year. With the exception of treatment with triple superphosphate $[Ca(H_2 PO_4)_2]$, tubers from plots with a preplant strip application of six Ca sources were less susceptible to bacterial soft rot than tubers from control plots. Tubers from plots treated with sieved $CaSO_4$ were again relatively higher in Ca concentration than those from other treatments (78,80).

The decreased susceptibility of tubers to bacterial soft rot could not be explained solely on the basis of increased concentrations of Ca in tubers, however. Thus, in the third experiment, application of triple superphosphate increased the Ca concentration in peel tissues of potato tubers, but the percent SAD was not significantly reduced from that of control plots. Obviously, other factors also may influence the resistance of tubers to bacterial soft rot in a given tuber sample.

Although the effectiveness of different Ca sources in raising tuber Ca concentration and in reducing tuber bacterial soft rot susceptibility in a given tuber sample varied from year to year, preplant strip application of sieved $CaSO_4$ gave very consistent and positive results in both regards (78,80). Furthermore, sieved $CaSO_4$ was also one of the best Ca sources for improving tuber grade and quality (72,73). On the other hand, application of lime was not effective in increasing Ca concentration in potato tubers and the quality of tubers was lower in comparison with other treatments (73).

Effectiveness of Calcium Application Methods

The method of field application of Ca whether applied broadcast, as a preplant strip, or as a sidedress, will influence the uptake of the cation by the potato tuber. Krauss and Marschner (51) obtained evidence that Ca can move directly into tubers from the soil through the epidermis. Furthermore, Kratzke and Palta (49,50) showed that stolon and tuber roots also play a role in uptake of water and calcium by tubers. Their studies with a divided pot system showed that addition of Ca to the stolon and tuber region resulted in a 3-fold increase in the Ca concentration of the tuber, whereas addition to the basal root region did not increase Ca above that of field-grown tubers. Applications of solutions of $CaCl_2$ and $Ca(NO_3)_2$ to plants at the time of maximal tuber growth were also highly effective in increasing tuber Ca concentrations (48). The time of application of $CaCl_2$ solutions for maximal uptake of Ca by tubers varied with the cultivar (48). Therefore, tuber Ca concentration would be strongly influenced by Ca concentrations in the soil surrounding the tuber as it develops. Since the preplant strip application method places Ca in a band in the hill where tubers are formed, it is very likely that this is one effective means of applying Ca to potato plants for tuber development.

In another field experiment, sieved $CaSO_4$ was applied at a rate of 336 kg of Ca/ha either broadcast, in a preplant strip, or as a sidedress. Tubers from all treated plots were less susceptible to bacterial soft rot, relative to those in control plots, regardless of the method of application. Although the Ca concentration in peel tissues of tubers receiving the preplant strip application of this mineral was higher than the concentration resulting from a sidedress application, tubers from all three treatments did not differ significantly with respect to the severity of surface area decayed.

Results with Ca application may differ, however, under differing environmental conditions. During the filling stage, tubers receive the greatest amount of their nutrients through the phloem (18). In contrast to other cations Ca moves throughout the plant almost exclusively in the xylem and preferentially to transpiring tissues. Thus, environmental factors such as high air temperatures and low humidity, which increase transpiration, will result in movement of greater amounts of Ca into the foliage than will less extreme conditions. If tubers have formed on plants, such factors also can result in movement of water out of tubers. With a return to low temperatures at night, high humidity, and ample soil moisture, the water deficit in the tuber can be restored with the solution comparatively high in Ca being carried directly from the stolon or tuber roots through the xylem. Cycles of such extreme conditions, which are more common in western states than Wisconsin, may cause the tuber alternately to expand and contract, resulting in an increase in Ca during each cycle.

INFLUENCE OF SOIL TYPE

Four sites were selected by Simmons et al. (72) to study the effectiveness of Ca applications on different soils in increasing Ca concentration in potato tubers. Tubers from these plots were examined for susceptibility to bacterial soft rot (78,80). The soils at these four sites represented the types present in the major potato growing regions of Wisconsin. They provided a range of soil conditions as follows: low cation exchange capacity (CEC), low exchangeable Ca and loamy sand soil at Hancock and Plover; medium CEC, high exchangeable Ca and silt loam soil at Antigo; and intermediate CEC with medium to high soil exchangeable Ca levels and sandy loam at Spooner. Five rates of sieved $CaSO_4$ were applied by a preplant application method ranging from 84 to 588 kg/ha Ca.

At the Hancock and Plover sites, tubers from Ca-fertilized plots developed less bacterial soft rot than tubers from the control plots. Calcium concentrations in peel tissues were increased from 0.132% to 0.244% DW, and from 0.088% to 0.147%, respectively, with a Ca application of 588 kg/ha. The correlation coefficients between Ca concentration in peel and percent SAD for tubers from these two locations were $r = -0.493$ ($P<0.05$) and $r = -0.531$ ($P < 0.01$), respectively. Application of Ca only slightly increased Ca concentrations in tubers grown at Antigo where soil Ca was relatively high; however, at Spooner where soil Ca was almost 200% higher than that at Hancock and Plover, Ca concentrations in tubers were not increased significantly and their susceptibility to bacterial soft rot was not reduced even at application rates of 588 kg/ha of Ca.

The above study indicated that different soil types influence the effectiveness of field applications of Ca. Simmons and Kelling (72) also showed that tuber yield was increased and grade was improved consistently with Ca fertilization on the low Ca loamy sand soils at Hancock and Plover, whereas results on the higher Ca soils at Antigo and Spooner were less consistent. In field trials as well as in laboratory experiments, the reduction in bacterial soft rot was highly correlated with the Ca increase in both peel and medullar tissues of the tuber, however. An increase in Ca in one tissue was related to the increase in the other. Direct movement of Ca into the tuber from the soil either directly through the periderm or through stolon and tuber roots (48,49,50), however, may have helped enrich the peel and could have helped account for the reduced gradient toward the center of the tuber (3,11,51).

Injection of Bacterial Cells and Pectic Enzyme Preparations into Tubers

Sets of potato tubers comprising a wide range of tuber Ca concentrations were obtained under controlled conditions in which plants were grown in sand supplied with 2 to 500 mg of Ca/L in nutrient solutions. Tuber Ca ranged from 0.014 to 0.062% DW, and from 0.031 to 0.115% DW, in medullar tissue of the cultivars Superior and Russet Burbank respectively. Calcium in the peel also was consistently lower in Superior tubers, from 0.026 to 0.283%, than it was in Russet Burbank tubers which contained 0.047 to 0.332%. Tubers of Superior and Russet Burbank with high and low medullar concentrations of Ca were injected with serial dilutions of Eca. After 96 hr in a nitrogen atmosphere, the diameter of decay at each injection site was measured. Decay diameters were consistently greater in low Ca tubers of both cultivars than in those with high levels of medullar Ca.

Strains of soft rot *Erwinia* produce both hydrolytic and lyase-type pectic enzymes, i.e. polygalacturonase (pg) and pectate lyase (PL), respectively, which are considered to have a primary role in tissue maceration (7,20,21,24,47). When sterile, partially purified culture filtrates of Eca containing Pl and PG were injected into surface-disinfested whole Russet Burbank tubers which were then incubated under anaerobic conditions, maceration was again more extensive in low Ca tissue than in high Ca tissue (58). Similar results obtained with inoculations of cells of Eca and injection of a sterile pectolytic preparation from Eca provides evidence for a major role that Ca may play in reduction of disease severity, i.e. interference with the action of pectic enzymes in tissue maceration (8,13,14).

Electrolyte Loss from Tuber Slices in Pectolytic Enzyme Preparations

The susceptibility of potato tubers to bacterial soft rot also has been associated with tissue membrane permeability and electrolyte leakage (40,83). An increase in cell membrane permeability can increase leakage of substrates that favor bacterial multiplication and subsequent soft rot development.

A considerable amount of the Ca in a plant is localized in cell walls; its concentration in this region is approximately 10^{-3} M (18,31). Calcium, being a divalent cation, has the ability to bridge two galacturonates via their carboxylate groups and Ca pectate is a principal component of the middle lamellae of plant cell walls in parenchymatous tissue (31). In addition, binding of proteins to polysaccharides through phenolic acid and Ca bridges has a strengthening influence on cell walls. The mechanical strength of the cell wall, determined by tests of cell wall extensibility,was increased by the addition of Ca (48). Calcium also binds anionic groups of all membranes to form bridges between structural components, thereby maintaining selective permeability, structural integrity, and cellular compartmentalization (10,67,70).

Stress induced leakiness of cells, whether caused by oxygen deficiency or freezing injury (48), also can be prevented or reversed by Ca treatments, and respiration of tubers infiltrated with Ca is greatly reduced over non-infiltrated tubers (3). Perhaps equally important, Ca appears to bind constituents of the plasmalemma loosely to the cell wall (5). The importance of Ca in a broad range of physiological functions (39,67,85) indicates that, in addition to effects on pectolytic enzyme activity, it may play several other roles in affecting the reaction of tubers to infection by soft rot Erwinias.

Tissue disks from Superior and Russet Burbank tubers were immersed in a partially-purified culture filtrate of Eca containing PL and PG. Measurements were made to determine the loss of electrolytes that accompanies the injury to cell membranes and cell lysis induced by the action of pectolytic enzymes (58). The Russet Burbank tubers with high Ca content used in these experiments contained 0.332% Ca in the peel and 0.115% in medullar tissue; low Ca tubers had concentrations of 0.117% and 0.037%, respectively. The high Ca Superior tubers had peel and medullar Ca concentrations of 0.283% and 0.062% Ca, respectively, whereas low Ca tubers contained 0.026% and 0.014%, respectively. When compared with an autoclaved, inactivated enzyme preparation, the rate of electrolyte loss in both cultivars was higher from peel and medullar tissues of the low Ca tubers than it was from high Ca tubers. The evidence of reduced electrolyte leakage from high Ca tissues also observed by Kratzke (48) indicates, in part, improved stability of plasma membranes which, in turn, can be related to Ca bridging of membrane components.

Degradation of Isolated Potato Cell Walls by Purified Pectate Lyase

Increase in the concentration of Ca in nutrient solutions supplied to potato plants of the cultivar Russet Burbank resulted in tubers with increased concentrations of Ca (78,79). Medullar tissue from tubers of plants watered over 80 days with Hoagland's solution and 250 ppm Ca contained 0.058% DW of the cation, whereas medullar tissue from plants lacking Ca supplementation had a concentration of 0.011%. Cell walls isolated from these tissues also differed in their Ca content. Without Ca supplementation, wall preparations contained 0.012% DW, Ca; with 250 ppm Ca in the nutrient solution, the concentration was increased to 0.076%.

High and low Ca cell walls were assayed for resistance to degradation by a purified pectate lyase (PLI) produced by a strain of Ecc (55). The rate of release of unsaturated uronides was significantly higher in low Ca walls. This study further supported the concept that enhanced resistance to bacterial soft rot in tubers with a high Ca content was related to the increased resistance to degradation by pectate lyase of high vs low Ca cell walls.

ROLE OF CALCIUM IN REDUCING TISSUE MACERATION BY SOFT ROT BACTERIA

Since Ca improves the structural integrity of both cell wall components and cell membranes, the reduction in the severity of Erwinia soft rot in high Ca potato tubers may be explained as a reduced rate of maceration resulting from enhancement of structural integrity. As noted previously, soft rot Erwinia macerate host tissues primarily by the action of pectolytic enzymes. Purified pectic enzymes acting on the pectic components of the middle lamella and cell wall can produce the same effects that characterize infections by pectolytic strains of Erwinia, i.e. maceration of tissue, increased leakage of electrolytes and death of cells (7,20,21,24).

Calcium ions may affect the degradation of pectic substances by either inhibiting or stimulating the pectic enzymes depending on the nature of the enzyme and the concentration of Ca (4,13,36,69,74). Calcium inhibits PG activity at relatively low concentrations. This observation is supported by data indicating that cation chelating agents such as ethylenediaminetetraacetic acid (EDTA) or oxalic acid enhance the action of PG on pectic substrates (7,13). In contrast, PL is stimulated by $CaCl_2$ at low concentrations ($2x10^{-5}$ to $1x10^{-3}$ M). However, Ca concentrations above 10^{-3}M *in vitro* result in precipitation of the Na-polypectate substrate and a decrease in the reaction rate of PL.

Potato disks were used by a number of researchers to study the effect of Ca and other cations on the maceration of plant tissues by pectolytic enzymes of fungi (8,36,58). Application of Ca at 10^{-4} M to reaction mixtures increased maceration of potato disks by PL produced by *Hypomyces solani* Reinke and Berth. f. sp. *cucurbitae* Synder and Hansen, but maceration was reduced at 10^{-3} M. A ratio of uronic acid to Ca less than or equal to 2 resulted in inhibition of PL from *Hypomyces* (36). These findings further support the concept that a reduction in bacterial soft rot severity is related to Ca effects on substrates of pectolytic enzymes in plant tissue.

Recently, gene cloning techniques have facilitated the evaluation of the roles of pectic enzymes in pathogenesis. *Escherichia coli* clones, expressing PL or PG genes from *Erwinia*, alone are able to cause localized maceration on potato tissues (20,21). However, adequate tests for evaluation of pathogenicity of these genetically engineered strains under natural conditions have not been developed. Furthermore, maceration of tissue disks or slices may not provide the most appropriate means for determination of pathogenicity. Other enzymes such as proteases, hemicellulases and cellulases may also be involved.

The possibility exists that increases in concentrations of Mg also can affect resistance of tuber tissue to maceration. Infiltration of tubers with $Mg(NO_3)_2$ reduced severity of decay, although not to the same degree as infiltration with $Ca(NO_3)_2$ (59). On the basis of studies on factors affecting resistance of tubers to gangrene caused by *Phoma exigua* Desm. var. *foveata* (Foister) Boerema (61), Olsson suggested that the lowered infection in high Mg tubers could be attributed in part to the linkage of this cation with pectic substances in the cell wall resulting in increased resistance to enzymatic degradation; however, our field trials indicated that applications with Mg resulted in a decrease in Ca content but higher concentrations of Mg, and this was associated with a decrease in resistance to bacterial soft rot (58). Pagel and Heitefuss (63) also concluded that the Mg content of tubers was not well correlated with susceptibility to bacterial soft rot.

It is recognized, however, that other physical and physiological relationships exist, indicating additional roles for Ca that could influence resistance (39,70). For example, Ca fertilization resulted in an increase in the netting of potato tuber surfaces (78). A higher percentage of tubers with well-developed netting was obtained from Ca fertilized plots (47.0 to 56.0%) than from control plots (20.5%) that did not receive Ca. Tubers with well-developed netting had more layers of periderm cells (average=13.6) than those tubers with smooth skin (average=9.1). Furthermore, periderm rupture force (measured by an Instron Universal Testing Instrument) for tubers with well-developed netting was significantly higher than for smooth-skinned low-calcium tubers. Presumably, increased thickness of tuber periderm may be related to the lower soft rot readings of high Ca tubers since increased periderm thickness may increase resistance to bruising and decrease pathways for invasion by soft rot bacteria.

SUMMARY AND CONCLUSIONS

Erwinia soft rot appears to be a more severe post-harvest problem in tubers produced in sandy soils low in cation exchange capacity (CEC) than in tubers from loamy soils. Tubers grown in soils with low CEC usually have lower concentrations of Ca than tubers from soils of high CEC, although the former may contain an adequate supply of Ca for normal growth of stems and leaves.

Cultivars of potato differ in concentrations of Ca in tubers (57,78). A direct correlation between tuber resistance of different cultivars and Ca concentrations in the tuber Ca has not been demonstrated, however. Highly susceptible cultivars usually are low in Ca, but tubers of the most resistant cultivars may not have the highest Ca concentration (1,57,63,78).

Increasing Ca levels in low CEC, low Ca soils results in an increase in Ca concentrations in potato tubers; furthermore, such tubers are more resistant to bacterial soft rot than tubers from soils not supplemented with Ca. In addition, increasing Ca content results in significant increases in the quality of tubers (72,73).

As noted previously, a number of additional factors also influence resistance of potato tubers to bacterial soft rot. Consideration must be given to these variables in attempts to evaluate effects of single factors such as Ca content on bacterial soft rot development. One of the most important of these other factors is oxygen status. Under low oxygen conditions, the tuber is very susceptible to tissue maceration by soft rot bacteria; whereas at atmospheric oxygen levels, very high populations of bacteria are required to initiate an infection, and lesions usually are localized (30,56). Potato tubers injected with a partially-purified pectolytic enzyme preparation from Ecc also decay readily at injection sites under low oxygen conditions, but tissues are highly resistant to enzymatic maceration under aerobic conditions (56).

The possibility exists that increases in soil Ca may, in some instances, have an adverse effect on potato tuber yield and quality. The specific affects of various soil treatments such as increased Ca levels upon the soil microflora are not known. Increases in soil Ca could enhance development of diseases caused by some soilborne pathogen, such as potato scab caused by *Streptomyces scabies* (Thaxter) Waksman and Henrici (29,42). Although researchers disagree as to the specific influence of nutritional factors on potato scab, many studies have related increases in soil Ca with increased disease severity. No obvious increases in potato scab were observed on high Ca tubers in any of our field trials, however.

Calcium fertilization is not a method for total control of *Erwinia* soft rot, but increasing this cation in tubers does result in a decrease in disease severity. The degree of disease reduction naturally also is under the influence of environmental and internal factors that affect tuber susceptibility as well as on the proper balance between Ca and Mg in fertilizing potatoes in low CEC soils and the cultivar used. Additional studies are needed on the form, manner, and the timing of Ca applications. The evidence is strong that presence of a readily available form of Ca in the tuber zone is important during time of maximal tuber bulking (48).

It is possible that proper application of Ca can not only reduce severity of *Erwinia* soft rot of potatoes, but also susceptibility to other diseases as well; in particular those diseases caused by pathogens that macerate tissues primarily with pectolytic enzymes (8,26,27,36). Use of Ca in reducing bacterial decay warrants additional study not only on potatoes, but on other fleshy vegetable and fruit crops and on certain ornamentals.

ACKNOWLEDGEMENTS

Sincere appreciation is expressed to Karen Simmons, Paul Fixen and Keith Kelling, Department of Soil Science, UW-Madison, for the design and completion of field trials and general advice and assistance in these studies. Support for the research was provided by the International Potato Center, Lima, Peru, the Wilson Geo. Meyer Co., and the College of Agricultural and Life Sciences, University of Wisconsin-Madison. We also thank Steven A. Vicen for preparation of the figures.

LITERATURE CITED

1. Abdel-Aal, S.A., Sellam, M.A., and Rushdi, M.H. 1987. Relation of chemical composition of certain potato varieties to their susceptibility to bacterial soft rot. pp. 140-141 Tenth Triennial Conf. European Assoc. Potato Res., Aalborg, Denmark.
2. Alberghina, A., Mazzucchi, V., and Pupillo, P. 1973. On the effect of an endopolygalacturonate trans-eliminase on potato tissues: The influence of water potential. Phytopath. Z. 78;204-213.
3. Arteca, R. 1982. Infiltration and movement of ^{45}Ca into potatoes. HortScience 17:757-758.
4. Atallah, M.T., and Nagel, C.W. 1977. The role of calcium ions in the activity of an endo-pectic-acid lyase on oligogalacturonides. J. Food Biochem. 1:185-206.
5. Atkinson, M.M., Baker, C.J., and Collmer, A. 1986. Transient activation of plasmalemma K^+ efflux and h^+ influx in tobacco by a pectate lyase isozyme from *Erwinia chrysanthemi*. Plant Physiol. 82:142-146.
6. Bangerth, F. 1979. Ca-related physiological disorders of plants. Ann. Rev. Phytopathol. 17:97-122.
7. Bateman, D.F., and Basham, H.G. 1976. Degradation of plant cell walls and membranes by microbial enzymes. Pages 316-355, in: Encyclopedia of Plant Physiology, Physiological Plant Pathology. Vol. 4. R. Heitefuss and P. H. Williams, Eds. Springer-Verlag, New York.
8. Bateman, D.F., and Lumsden, R.D. 1965. Relation of calcium

content and nature of pectic substances in bean hypocotyls of different ages to susceptibility to an isolate of *Rhizoctonia solani*. Phytopathology 55: 734-738.

9. Biehn, W.L., Sands, D.C., and Hankin, L. 1972. Relationship between percent dry matter content of potato tubers and susceptibility to bacterial soft rot. Phytopathology 62:747 (Abstr.)

10. Blowers, D.P., Boss, W.F., and Trewavas, A.J. 1988. Rapid changes in plasma membrane protein phosphorylation during initiation of cell wall digestion. Plant Physiol. 86:505-509.

11. Bretzloff, C.W., and McMenamin, J. 1971. Some aspects of potato appearance and texture. III. Sampling tubers for cation analysis. Am. Potato J. 48:97-104.

12. Buelow, F., Maher, E.A., and Kelman, A. 1986. Assessment of bacterial soft rot potential. Pages 440-455, in: Engineering for potatoes. B. F. Cargill, Ed. Michigan State Univ. and Am. Soc. Agric. Engin., St. Joseph, MI.

13. Buescher, R.W., Hudson, J.M., and Adams, J.R. 1979. Inhibition of poly-galacturonase softening of cucumber pickles by calcium chloride. J. Food Sci. 44:1786-1787.

14. Buescher, R.W., and Hobson, G.E. 1982. Role of calcium and chelating agents in regulating the degradation of tomato fruit tissue by polygalacturonase. J. Food Biochem. 6:147-160.

15. Burton, W., and Wigginton, M.J. 1970. The effect of a film of water on the oxygen status of a potato tuber. Potato Res. 13:180-186.

16. Cappellini, R.A., Ceponis, M.J., Wells, J.M., and Lightner, G.W. 1984. Disorders in potato shipments to the New York market, 1972-1980. Plant Dis. 68:1018-1020.

17. Christiansen, M.H., and Foy, C.D. 1979. Fate and function of calcium in tissue. Commun. Soil Sci. Plant Anal. 10:427-442.

18. Clarkson, D.T. 1984. Calcium transport between tissues and its distribution in the plant. Plant, Cell and Envir. 7:449-456.

19. Collier, G.F., Wurr, D.C.E., and Huntington, V.C. 1978. The effect of calcium nutrition on the incidence of internal rust spot in the potato. J. Agric. Sci. (Cambridge) 91:241-243.

20. Collmer, A. 1987. Pectic enzymes and bacterial invasion of plants. Pages 253-284 in: Plant Microbe Interactions. T. Kosuge and E. Nester, Eds. MacMillan, New York.

21. Collmer, A., and Keen, N.T. 1986. The role of pectic enzymes in pathogenesis. Ann. Rev. Phytopathol. 24:383-409.

22. Conway, W.S., Gross, K.C., and Sams, C.E. 1987. Relationship of bound calcium and inoculum concentration to the effect of

postharvest calcium treatment on decay of apples by *Penicillium expansum*. Plant Dis. 71:78-80.

23. Conway, W.S., Gross, K.C., Boyer, C.D., and Sams, C.D. 1988 Inhibition of *Penicillium expansum* polygalacturonase activity by increased apple cell wall Ca. Phytopathology 78:1052-1055.

24. Cooper, R.M. 1983. The mechanisms and significance of enzymic degradation of host cell walls by parasites. Pages 102-135 in: Biochemical Plant Pathology. J. A. Callow, Ed. John Wiley, England.

25. Cother, E.J., and Cullis, B.R. 1987. Seed tuber susceptibility to *Erwinia chrysanthemi*: evaluation of altered tuber physiology as a means of reducing incidence and severity of soft rot. Potato Res. 30:229-240.

26. Csinos, A.S., and Gains, T.P. 1986. Peanut pod rot complex: a geocarposphere nutrient imbalance. Plant Dis. 70:525-529.

27. Csinos, A.S., and Gains, T.P., and Walker, M.E. 1984. Involvement of nutrition and fungi in the peanut pod rot complex. Plant Dis. 68:61-65.

28. Davies, H.V., and Millard, P. 1985. Fractionation and distribution of calcium in sprouting and non-sprouting potato tubers. Ann. Bot. 56:745-754.

29. Davis, J.R., M^CDole, R.E., and Callihan, R.H. 1976. Fertilizer effects on common scab of potato and the relation of calcium and phosphate-phosphorus. Phytopathology 66:1236-1241.

30. DeBoer, S.H., and Kelman, A. 1978. Influence of oxygen concentration and storage factors on susceptibility of potato tubers to bacterial soft rot (*Erwinia carotovora*). Potato Res. 21:65-80.

31. Demarty, M., Morvan, C., and Thellier, M. 1984. Ca and the cell wall. Plant, Cell Envir. 7:441-448.

32. Dunn, H.V., and Rost, C.O. 1948. Effect of fertilizer on the composition of potatoes grown in the Red River Valley of Minnesota. Soil Sci. Soc. Amer. Proc. 13:374-379.

33. Dyson, P.W., and Digby, J. 1975. Effects of calcium on sprout growth and sub-apical necrosis in Majestic potatoes. Potato Res. 18:290-305.

34 Edgington, L.V., Corden, M.E., and Dimond, A.E. 1961. The role of pectic substances in chemically-induced resistance to wilt of tomato. Phytopathology 51:179-182.

35. Fritz, V.A., Honma, S., and Widders, I. 1988. Effects of petiole calcium status, petiole location, and plant age on the incidence and progression of soft rot in Chinese cabbage. J. Amer. Soc. Hort. Sci. 113:56-61.

36. Hancock, J.G., and Stanghellini, M.E. 1968. Calcium localization

in *Hypomyces*-infected squash hypocotyls and effect of Ca on pectate lyase activity and tissue maceration. Can. J. Bot. 46:405-409.

37. Geraldson, C.M. 1954. The control of blackheart of celery. Proc. Am. Soc. Hort. Sci. 63:353-358.

38. Geraldson, C.M. 1957. Control of blossom-end rot of tomatoes. Proc. Am. Soc. Hort. Sci. 69:309-317.

39. Hepler, P.K., and Wayne, R.O. 1985. Calcium and plant development. Ann. Rev. Plant Phys. 36:397-439.

40. Hidalgo, O.A., and Echandi, E. 1983. Influence of temperature and length of storage on resistance of potato to tuber rot induced by *Erwinia crysanthemi*. Am. Potato J. 60:1-18.

41. Hiller, L.K., Koller, D.C., and Thornton, R.E. 1985. Physiological disorders of potato tubers. Pages 389-455, in: Potato Physiology. P. H. Li, Ed. Academic Press, Inc., N.Y.

42. Horsfall, J.G., Hollis, J.P., and Jacobson, H.G.M. 1954. Calcium and potato scab. Phytopathology 44:19-24.

43. Huber, D.M. 1981. The use of fertilizers and organic amendments in the control of plant disease. Pages 357-394, in: Handbook of pest management in agriculture. Vol. I. D. Pimentel, Ed. CRC Press, Inc., Florida.

44. Kelman, A., Baughn, J.W., and Maher, E.A. 1978. The relation of bacterial soft rot susceptibility to water status of potato tubers. Phytopathol. News 12:178. (Abstr.).

45. Kirkby, E.A., and Pilbeam, D.J. 1984. Calcium as a plant nutrient Plant, Cell Envir. 7:397-405.

46. Konno, H., Yamaya, T., Yamasaki, Y., and Matsumoto, H. 1984. Pectic polysaccharide breakdown of cell walls in cucumber roots grown with calcium starvation. Plant Physiol. 76:633-637.

47. Kotoujansky, A. 1987. Molecular genetics of pathogenesis by soft rot *Erwinias*. Ann. Rev. Phytopathol. 25:405-430.

48. Kratzke, M.G. 1988. Study of mechanism of calcium uptake by potato tubers and cellular properties affecting soft rot. Ph. D. Thesis. Univ. of Wisconsin-Madison. 321p.

49. Kratzke, M.G., and Palta, J.P. 1985. Evidence for the existence of functional roots on potato tubers and stolons: significance in water transport to the tuber. Am. Potato J. 62:227-236.

50. Kratzke, M.G., and Palta, J.P. 1986. Calcium accumulation in potato tubers: role of the basal roots. Hort. Science 21:1022-1024.

51. Kraus, A., and Marschner, H. 1971. Influence of direct supply of calcium to potato tubers on the yield and the calcium content. Z. Pflanzenernahr. Bodenkd. 129:1-9.

52. Lovrekovich, L., Lovrekovich, H., and Stahmann, M.A. 1967.

Inhibition of phenol oxidation by *Erwinia carotovora* in potato tuber tissue and its significance in disease resistance. Phytopathology 57:737-74.

53. Lund, B.M., and Kelman, A. 1977. Determination of the potential for development of bacterial soft rot of tubers. Am. Potato J. 54:211-225.

54. Lyon, G.D., Lund, B.M., Bayliss, G.E., and Wyatt, G.M. 1975. Resistance of potato tubers to *Erwinia carotovora* and formation of rishitin and phytuberin in infected tissue. Physiol. Plant Pathol. 6:43-50.

55. Maher, E.A., Livingston, R.S., and Kelman, A. 1986. Recognition of pectate lyase in western blots by monoclonal antibodies. Phytopathology 76:1101. (Abstr.).

56. Maher, E.A., and Kelman, A. 1983. Oxygen status effects on maceration of potato tuber tissue by pectic enzymes produced by *Erwinia carotovora*. Phytopathology 73:536-539.

57. McGuire, R.G., and Kelman, A. 1983. Susceptibility of potato cultivars to *Erwinia* soft rot. Phytopathology 73:809. (Abstr.)

58. McGuire, R.G., and Kelman, A. 1984. Reduced severity of *Erwinia* soft rot in potato tubers with increased Ca content. Phytopathology 74:1250-1256.

59. McGuire, R.G., and Kelman, A. 1986. Calcium in potato tuber cell walls in relation to tissue maceration by *Erwinia carotovora* pv. *atroseptica*. Phytopathology 76:401-406.

60. Myers, D.F., and Campbell, R.N. 1985. Lime and the control of clubroot of crucifers: Effects of pH, calcium, magnesium and their interactions. Phytopathology 75:670-673.

61. Olsson, K. 1984. Some biochemical aspects of resistance to potato gangrene. Pages 79-80, in: 9th Triennial Conf. European Assoc. Potato Res. F. A. Winiger and A. Stokli, Eds. Interlaken, Switzerland.

62. Otzau, V., and Secor, G.A. 1981. Soft rot susceptibility of potatoes with high reducing sugar content. Phytopathology 71:290-295.

63. Pagel, W., and Heitefuss, R. 1987. Investigations on the role of calcium and some cell wall properties for the susceptibility of potato cultivars against *Erwinia carotovora*. Pages 141-142, in Tenth Triennial Conf. European Assoc. Potato Res. N. E. Foldo, S. E. Hansen, N. K. Nielsen, and R. Rasmussen, Eds. Aalborg, Denmark.

64. Park, S.K. 1969. Studies on the relationship between calcium nutrient and soft rot disease in Chinese cabbage. The research report of the Office of Rural Development. Suroon, Korea 12:63-

70.

65. Perombelon, M.C M., and Kelman, A. 1980. Ecology of the soft rot *Erwinias*. Ann. Rev. Phytopathol. 18:361-387.

66. Platero, M., and Tejerina, G. 1976. Calcium nutrition in *Phaseolus vulgaris* in relation to its resistance to *Erwinia carotovora*. Phytopathol. Z. 85:314-319.

67. Pooviah, B.W., and Reddy, A.S.N. 1987. Calcium messenger system in plants. CRC Critical Reviews in Plant Sciences. 6:47-103.

68. Power, R.H. 1983. Relationship between the soil environment and tomato resistance to bacterial wilt (*Pseudomonas solanacearum*): 4. Control Method. Surinam Agric. 31:39-47.

69. Pratt, A.L., and McIntyre, G.A. 1972. Effects of some divalent cations on the macerating activity of two pectic lyases produced by *Pseudomonas fluorescens*. Phytopathology 62:499. (Abstr.).

70. Roux S.J., and Slocum, R.C. 1982. Role of calcium in mediating cellular functions important for growth and development in higher plants. Pages 409-451, in: Calcium and cell function. W. Y. Cheung, Ed. Academic Press, New York.

71. Sellam, M.A., Rushdi, M.H., and Abdel-Aal, S.A. 1980. Relation of chemical composition of certain potato varieties to their susceptibility to bacterial soft rot. Egypt. J. Phytopathol. 12:137-143.

72. Simmons, K.E., and Kelling, K.A. 1987. Potato responses to calcium application on several soil types. Am. Potato J. 64:119-136.

73. Simmons, K.E., Kelling, K.A., Wolkowski, R.P., and Kelman, A. 1988. Effect of calcium source and application method on potato yield and cation composition. Agron. J. 80:73-21.

74. Starr, M.P., and Moran, F. 1962. Eliminative split of pectate substances by phytopathogenic soft rot bacteria. Science 135:920-921.

75. Sun, S.K., and Huang, J.W. 1985. Formulated soil amendment for controlling Fusarium wilt and other soilborne diseases. Plant Dis. 69:917-920.

76. Tripathi, R.K., and Verma, M.N. 1975. Phenolic compounds and polyphenol oxidase activity in relation to resistance in potatoes against bacterial soft rot. Indian J. Exp. Biol. 13:414-416.

77. True, R.H., Hogan, J.M., Augustin, J., Johnson, S.R., Teitzel, C., Toma, R.B., and Shawn, R.L. 1978. Mineral composition of freshly harvested potatoes. Am. Potato J. 55:511-519.

78. Tzeng, K. C. 1986. Calcium nutrition of potato plants in relation to bacterial soft rot susceptibility and internal brown spot of

potato tubers. Ph.D. Thesis. University of Wisconsin-Madison. 150 p.

79. Tzeng, K.C., Kelman, A., Simmons, K.E., and Kelling, K.A. 1986. Relationship of calcium nutrition to internal brown spot of potato tubers and sub-apical necrosis of sprouts. Am. Potato J. 63:87-97.

80. Tzeng, K.C., Kelman, A., Simmons, K.E., and Kelling, K.A. 1985. Relation of calcium content in potato tuber periderm to bacterial soft rot susceptibility, internal brown spot and sub-apical necrosis on sprouts. Phytopathology 75:1379 (Abstr).

81. Weber, J. 1983. The role of pectin on the significance of varietal and seasonal differences in soft rot susceptibility of potato tubers. Phytopathol. Z. 108:135-142.

82. Weintraub, M., and Ragetli, H.W.J. 1961. Cell Wall composition of leaves with a localized virus infection. Phytopathology 51:215-219.

83. Workman, M., Kerschner, E., and Harrison, M. 1976. The effect of storage factors on membrane permeability and sugar content of potatoes and decay by *Erwinia carotovora* var. *atroseptica* and *Fusarium roseum* var. *sambucinum*. Am. Potato J. 53:191-204.

84. Workman, M., and Holm, D.G. 1984. Potato clone variation in blackspot and soft rot susceptibility, redox potential, ascorbic acid, dry matter and potassium. Am. Potato J. 61:723-733.

85. Zook, M., Rush, J.S., and Kuc, J. 1987. A role for calcium in the elicitation of rishitin and lubimin accumulation in potato tuber tissue. Plant Physiol. 84:520-525.

PATHOLOGY AND NUTRITION IN THE PEANUT POD ROT COMPLEX

Alex S. Csinos and Durham K. Bell
Department of Plant Pathology
University of Georgia
Coastal Plain Experiment Station
Tifton, GA 31793

Peanut pod rot (pod breakdown) is a sporadic but common disease of peanut (*Arachis hypogaea* L.) which causes serious losses throughout all peanut growing regions of the world (5,9,15,19,26,44). Symptoms of the disease vary depending on location (14,26,40), time of season (24) and incitants (26). Deterioration of fully formed pods occurs after pods develop either a tan, brown, dry, decay or a greasy, black, wet decay depending on causal organisms and prevailing environmental conditions (20,26). Often, many pods are left in the soil at digging since pegs are weakened or decayed at ground level.

There are no above ground symptoms of pod rot, other than that severely affected plants may be darker green, and flowering is prolonged more than plants with a normal crop of pods. Since the root systems are generally not infected, the reduced demand for carbohydrate by the loss of the fruit usually increases the vigor of the foliage. Plants with the greatest degree of pod rot near harvest will appear the most vigorous and provide no indication of serious disease losses below the soil surface.

The peanut is a unique legume since it flowers above ground and the pods (fruit) are formed below the soil surface. As a result of anatomical separation, few nutrients are translocated directly from roots to the pods. The pods are formed after pollination when the pegs (carpophores) elongate and enter the soil. Peanut pods develop horizontally and mature beneath the soil surface (1-8 cm) over a period of ca. 9-12 wk. During this period, the pods are susceptible to numerous environmental, entomological and pathological problems.

PATHOLOGY

This overview of the peanut pod rot disease complex (24,25) focuses on *Pythium myriotylum* Drechs. and *Rhizoctonia solani* Kühn as incitants, plus certain other soil biota that affect the disease. The severity of disease caused by both of these organisms has been shown to be influenced markedly by plant nutrition, which will be discussed elsewhere in this chapter. We recognize, however, that other soilborne pathogens, including *Cylindrocladium crotalariae* (Loos) Bell and Sobers, *Sclerotinia minor* Jagger and *Sclerotium rolfsii* Sacc., often cause severe pod rot, with accompanying yield and quality reductions (40), but plant nutrition appears to have a minor role in determining plant susceptibility to these pathogens.

Frezzi (17) in Argentina was apparently the first to report that *Pythium* spp. were associated with peanut pod rot. He stated that *P. debaryanum* Hesse, *P. irregulare* Buis. and *P. ultimum* Trow caused decay of immature and mature pods.

Garren (20) in Virginia was among the first in the United States to determine that peanut pods (geocarps) have a complex rhizoplane-like (geocarposphere) microflora, that this microflora changes with moisture and increasing temperature as the growing season advances and that *P. myriotylum* and *R. solani* are two potential incitants of pod rot. Garren (21) reported subsequently that the ratios of these potential pathogens in decaying pods varied from ca. 1:1, to *P. myriotylum* greatly predominating over *R. solani*. He determined in this study that the soil fungicide PCNB (pentachloronitrobenzene) reduced the incidence of *R. solani*, but increased pod rot caused by *P. myriotylum*. This effect resulted in Garren suggesting and demonstrating that *R. solani* was antagonistic to *P. myriotylum* (27).

Garren (25) inoculated actively growing pods attached to the plant or detached pods in nonsterile soil by infesting soil separately with pure cultures of *P. myriotylum*, *R. solani* and *Zygorhynchus moelleri* Vuill. The results were 50 and 100% of pods inoculated with *P. myriotylum* decayed after 2 and 7 days, respectively. Thirty days after inoculation 16% of the attached pods inoculated with *Z. moelleri* and 31% of those inoculated with *R. solani* were decayed. All pods inoculated with *P. myriotylum* and incubated at 32 and 39C decayed in 2 days, again substantiating that pod rot caused by *P. myriotylum* was favored by high temperatures. When decaying pod tissue was plated on a selective medium for *Pythium* spp., *P. myriotylum* grew from 88% of pods inoculated with the fungus. However, *P. myriotylum* also grew from 15 to 35% of pods inoculated, respectively, with *R. solani* and *Z. moelleri*. Thus, *P. myriotylum* was considered to fulfill "Koch's

Postulates" as a primary pathogen of peanut pods; pathogenicity of *R. solani* and *Z. moelleri* was regarded as questionable. *Rhizoctonia solani* has remained as a potential pod rot pathogen, but *Z. moelleri* has ceased to be seriously considered in this role.

Frank (14) reported that peanut pod rot has been prevalent in Israel since 1959. *Pythium* spp. grew from diseased pods and the predominant species was *P. myriotylum*. The fruiting zone of peanut plants growing in field microplots was infested with *P. myriotylum*. Many pods infected with *Pythium* were partially or completely decayed. When soil under developing pods was infested, the pathogen spread and decayed most of the pods.

Garren (24) compiled and reported considerable evidence indicating that peanut pod rot was a disease complex. He conducted six tests in four fields during a four year period. The phycomycete-specific fungicide fenaminsulf [na-p-(dimethylamino) benzenediazo sulfonate] was used to control *P. myriotylum* and PCNB to control *R. solani*. The results with these fungi specific fungicides led Garren to propose "three theories on peanut pod rot in Virginia."

(a) "when inoculum potential of *P. myriotylum* is high and weather is moderately favorable to maintenance of this potential, *P. myriotylum* overcomes the natural resistance of developing peanut fruits and causes pod rot."

(b) "when inoculum potential of *R. solani* is high and weather is moderately favorable to maintenance of this potential, *R. solani* overcomes rot resistance of developing peanut fruits and causes pod rot."

(c) "weather conditions which usually prevail maintain a high inoculum potential of *P. myriotylum* but are unfavorable to *R. solani*."

Garren (24) also confirmed his concept that there existed in Virginia a "field capacity" for pod rot. That is, different fields, either in adjacent juxtaposition, near or far apart, could have varying intensities of pod rot simultaneously under nearly identical environmental conditions.

Garren (26) determined that *R. solani* caused pod rot similar to that incited by *P. myriotylum*, when the potting medium of 35-70 day old plants with attached pods was infested with *R. solani* and maintained at 20-28C. This temperature was considerably lower than the general optimum of 32+C for *P. myriotylum*. Pods were harvested from 150 day old plants and the amount of pod rot caused by both fungi compared. Garren observed that disease caused by *R. solani* developed more slowly and less extensively than that incited by *P. myriotylum*.

Ashworth and Langley (1) in Texas summarized 10 field experiments during 1960-1962 by reporting that ca. 87% of "pod damage" in Texas was caused by *R. solani* alone or combined with insect larvae. Other soil fauna have been associated with pod rot. Porter and Smith (39) in Virginia reported that field grown pods injured by southern corn rootworm (*Diabrotica undecimpunctata howardi* Barber) larvae were more susceptible to infection by various fungi than noninjured pods. With combined high inoculum concentrations of southern corn rootworm larvae and *P. myriotylum*, pod rot averaged 40% more than with *P. myriotylum* alone. Garcia and Mitchell (18) in Florida demonstrated that pods exposed to moderate, combined inoculum levels of the peanut root knot nematode [*Meloidogyne arenaria* (Neal) Chitwood] and *P. myriotylum* sustained 31% more decay than with P. myriotylum alone. Shew and Beute (44) in North Carolina stated that soil mites (*Caloglyphus* spp.) were associated with over 50% of decaying field grown pods where *P. myriotylum* was the primary fungal pathogen. *Pythium myriotylum* was chosen by ca. 98% of mites responding to a fungal-food preference selection, and the fungus grew from 90% of mite fecal pellets after feeding on *P. myriotylum*. Oospores of *P. myriotylum* were viable after passage through the alimentary canal of mites. In greenhouse and field tests, several acaricides and broad spectrum insecticides significantly reduced pod rot caused by *P. myriotylum*.

Fusarium spp., primarily *F. solani* (Mart.) Appel and Wr., have been implicated in the *Pythium-Rhizoctonia* pod rot complex. Kranz and Pucci (35) in Libya reported that pod rot there was caused largely by *F. solani* and other *Fusarium* spp. Garren (20,24,26) implicated *Fusarium* spp. in *Pythium-Rhizoctonia* pod rot in the United States and in one article (24) he stated that under Virginia field conditions *Fusarium* spp. preceded infection by *P. myriotylum*.

Frank (14) found that pod rot caused by *P. myriotylum* in field microplots containing heat-treated soil was less severe than with nontreated soil. He considered that the probable reason for this was an enhancement of virulence of *P. myriotylum* by frequent association of *Fusarium* spp. in nontreated soil. In heat-treated soil, *Fusarium* spp. were rarely associated with *Pythium* pod rot.

Frank (16) grew peanuts in field microplots and inoculated the pods by infesting the soil up to ca. the middle period of pod development. In heat-treated soil, *F. solani* alone caused 6% pod rot compared to 31% incited by *P. myriotylum* alone. Inoculation by *F. solani* initially followed by *P. myriotylum* caused 54% disease and the reverse order of inoculation produced 36%. Inoculation with both fungi simultaneously incited 39% decayed pods. He (16) then used *P.*

myriotylum and *F. solani* separately and the two combined, each at one- half inoculum concentration, in nontreated soil and obtained 43, 23, and 45%, respectively, disease. Based on these results, Frank (16) postulated a three step progression of pod rot in Israel. In Step 1, *F. solani* predisposed pods to sporadic infection by *P. myriotylum*. Step 2 involved pod colonization by P. myriotylum and the rapid increase of pod rot, and in Step 3, *F. solani* and saprophytic organisms caused the disintegration of pods, with a sharp reduction or disappearance of pathogenesis by *P. myriotylum*.

Garcia and Mitchell (19) tested in a greenhouse the pathogenic etiology in pod rot by exposing pods attached to the plant or detached pods to various fungi singly or in series. After exposure of pods to *R. solani*, the isolation frequency of *P. myriotylum* was reduced and rot induced by *Pythium* was suppressed or prevented. High infestation levels of *Macrophomina phaseolina* (Tassi) Goid counteracted the antagonistic effect of *R. solani* to *P. myriotylum* with pods grown in soil but not in vermiculite. Exposure of pods to *P. myriotylum* followed by *R. solani* greatly reduced or eliminated recovery of *R. solani*. The authors (18) demonstrated that combined inoculation of pods growing in tubes of soil with *P. myriotylum* and *F. solani* caused 28% more pod rot than *P. myriotylum* alone.

In summary, peanut pod rot as a disease complex researched by scientists in Argentina, the United States and Israel has similarities and differences. Perhaps the most obvious similarities are reports of *Pythium* spp., mainly *P. myriotylum*, as the primary pathogen in Argentina (17), Virginia (24,39), North Carolina (44), Florida (18), and Israel (16). The most obvious difference perhaps is that *R. solani* is indicated as the major pathogen in Texas (1), but is not mentioned as a component of the disease complex in Argentina (17), or Israel (16). *Rhizoctonia solani* was considered as a potentially severe pod rot pathogen in Virginia, but usually it was less virulent than *P. myriotylum*, due to prevailing environmental conditions (24,26).

Fusarium spp., mainly *F. solani* are reported to enhance *Pythium* pod rot in Virginia (20), Florida (18) and Israel (16). *Fusarium* spp. were reported as major pod rot pathogens in Libya (35).

Soil fauna are reported to increase *Pythium* pod rot in Virginia (39), North Carolina (44) and Florida (18), and *Rhizoctonia* pod rot in Texas (1).

NUTRITION

As early as 1954, Higgins in Georgia (33) described a condition of peanut he named black pod. The symptoms included a collapse of seed and pod tissues which were then attacked by various soil fungi causing the dark discoloration. Higgins attributed the condition to drought and deficiency of calcium in the soil. He also noted that the problem was more prominent on large seeded Virginia peanuts than on smaller seeded Runner and Spanish types.

Several years later, in 1964, Garren (20) described a pre-harvest fruit rot of peanut in Virginia which caused up to 32% of the fruit to decay. He indicated that *P. myriotylum* and *R. solani* were the principal "pod breakdown" pathogens (20,23,24,26).

However, research on the disorder uncovered an interesting finding. Relatively large amounts of gypsum ($CaSO_4.2H_2O$) applied to peanuts at bloom reduced the severity of pod breakdown (21). Gypsum at 2060 kg/ha was optimum for pod breakdown suppression and high fruit yield. However, the theory that calcium deficiency could somehow be the cause of pod breakdown appeared contradictory since 490 to 980 kg/ha of gypsum was applied as a common practice in Virginia and often soil levels of calcium were initially high.

The need for available Ca in the fruiting zone of peanut has been well documented (3,4,5,7,8,31,42,43,46,48,49,50). However, in the presence of available calcium, conditions which may interfere with calcium absorption by the fruit may result in aborted or abnormal fruit development (6,33,37,45).

In other studies Hallock and Garren (32) demonstrated that high rates of gypsum (1030-3090 kg/ha) applied to "Virginia Bunch 46-2" reduced pod breakdown 2 out of 3 years, increased pod yields and percent sound mature seeds. In all three years of the study, gypsum increased calcium in pods and decreased percent potassium. Applications of $MgSO_4$ (1345 kg/ha) and K_2SO_4 (1010-2020 kg/ha) increased pod breakdown, and percent potassium in pods, and decreased or tended to decrease fruit yield and calcium in pods. They indicated that the vulnerability to pod breakdown pathogens was reduced in pods containing 0.20% or more calcium.

The influence of plant nutrients on disease was suggested by Hallock and Garren (32) to be several fold. Polygalacturonase activity in *Rhizoctonia*-infected bean hypocotyls was greatly reduced in calcium or barium solutions, but occurred readily in potassium and sodium solutions (2). Calcium, an integral part of calcium pectate, affords resistance to polygalacturonase. Calcium may promote formation of dormant, less pathogenic growth stages, while K may promote the active growth stages of *Pythium* spp. (52). The disease enhancement of K_2SO_4 and $MgSO_4$ also may be the result of nutritional interference in Ca nutrition of the fruit.

Calcium nutrition in peanut has long been studied and is considered one of the most important nutritional aspects of peanut production in the U.S.A. Bledsoe et al. (3) in 1949 demonstrated the absorption of calcium by fruit using radioactive calcium. Their study showed that calcium absorption by the developing fruit represented the sole avenue of calcium nutrition of the pod. In a study where pods and roots were grown in separate containers radioactive ^{45}Ca applied to the peanut root system was not detected in shells or seed. Similarly, Walker (47) has demonstrated that calcium applied to foliage does not provide calcium nutrition to the fruit.

Skelton and Shear (45) demonstrated the lack of ^{45}Ca translocation into pods that are not transpiring; however, if pods were exposed to the atmosphere and allowed to transpire, ^{45}Ca was detected in pods. Intermittent drying of calcium-free pods did not result in an increased movement of calcium into the fruit from the vegetative part of the plant. The most critical period for calcium availability in the pod zone was found to be 15 to 35 days after the pegs (gynopores) reached the soil (7). Calcium is not translocated from one side of the plant to the other, but must be in an available form around all pods in the soil.

In 1974, Moore and Wills (38) studied the effect of gypsum on pathogenicity of isolates of *P. myriotylum* and *R. solani* to peanut pods. They found that gypsum added to artificial sterile medium did not alter the susceptibility of peanut fruit to either *P. myriotylum* or *R. solani* inoculated alone or in combination. Inoculum was added at the rate of 5g of a sand cornmeal culture to 140 cc of Weblite. This quantity of inoculum may be artificially high and may not represent typical field inoculum potentials. In addition, no attempt was made to correlate calcium uptake or calcium content of pods with susceptibility to pod rot.

In 1980, Walker and Csinos in Georgia (51) reported on a three-year study conducted in a location with a low soil calcium level (356 kg/ha) and in one location with a higher soil calcium level (752 kg/ha). Five peanut cultivars, Florunner, Tifrun, Florigiant, GA 194 VA, and Early Bunch were topdressed with 0, 560, 1120 or 1680 kg/ha of gypsum at early bloom (Ca. 45-50 days after planting). Pod rot did not occur on any cultivar in any treatment at the high calcium location. At the low calcium location severe pod rot occurred on plots receiving no gypsum, but the severity decreased for all cultivars as the rate of gypsum increased. Significant differences occurred among cultivars for pod rot and agronomic parameters. Cultivars with high calcium requirements (e.g. Early Bunch) were more susceptible to pod rot than cultivars less dependent on calcium fertilization (e.g. Florunner).

In 1984, Csinos et al. (9) reported on studies which evaluated fungi-specific-chemicals and fertilizers to determine their effects on pathogen involvement in the pod rot complex. Although *Pythium*, *Rhizoctonia*, and *Fusarium* spp. were isolated from soil and decaying pods and pegs throughout the growing season, no consistent differences were found among treatments for soil populations or isolations from decaying pods. Plots treated with a calcium source were generally higher in yield and grade and tended to be lower in incidence of pod rot. There was a significant positive correlation with most elements in fruits and pod rot except calcium which had a significant negative correlation to pod rot. They proposed that fungi are secondary to the disease complex and nutritional deficiency or imbalance may be the primary cause.

In 1986, Csinos and Gaines (10) reported that the peanut pod rot complex appeared to be initiated by a geocarposphere nutrient imbalance. The peanut fruiting zone treated at early bloom with fertilizers such as $MgSO_4$, NH_4NO_3 and K_2SO_4 tended to increase pod rot and decrease grade and yield, while gypsum decreased pod rot and increased grade and yield. In addition, the detrimental effects of $MgSO_4$, NH_4NO_3 or K_2SO_4 were ameliorated by a concomitant application of gypsum. The reduction in pod rot by the concomitant applications of gypsum with other fertilizers appeared to be due to the reductions of concentrations of magnesium and nitrogen in the pods compared to peanut treated with just $MgSO_4$ and K_2SO_4 or NH_4NO_3. Apparently, gypsum through an undescribed soil reaction affected potassium and magnesium removal from the pod development zone, thus explaining the lower potassium and magnesium levels in the geocarposphere soil and pods.

Jacobson et al. (34) studying the influence of calcium on selectivity of ion absorption, suggested that calcium affects a screening of ions at the cell surface, presumably nonmetabolic in nature, which is followed by a metabolic absorption step. In their studies, calcium drastically altered the ratio of absorption by several plant roots of sodium and potassium from a mixture of the two. They further suggest that the inhibitory effect of absorption of ions appears to preclude the role of calcium as a simple protective agent.

Striking similarities of the peanut pod rot complex and blossom-end rot of tomato and pepper are evident from the studies of Hallock and Garren (32), Boswell and Thames (5) and Csinos et al. (9,10). Blossom-end rot of fruits is characterized by a green water-soaked area at the blossom end of the fruit that turns gray to dark brown, then shrinks and collapses. This area is commonly invaded by microorganisms resulting in further fruit decomposition. Applications of calcium that resulted in increased fruit calcium, prevented blossom-end rot (12,28,36). In addition, Geraldson (29) reported that excessive soluble ammonium, potassium, magnesium and sodium salts or a deficiency of soluble calcium increased blossom-end rot. Raleigh and Chucka (41) indicated from a nutrient study on tomato that the ratio among elements is more important as a cause of blossom-end rot than the actual concentration of the elements. Gerard and Hipp (30) and DeKock et al. (11) concluded in their studies that climatic stress where calcium movement was limited by poor water relations in the plant was a major contributor to blossom-end rot of tomato.

The key factor in reduction of peanut pod rot or pod breakdown in the studies of Csinos et al. (9,10) and Hallock and Garren (21) has been the increased concentration of calcium in the peanut fruit. Boswell and Thames (5) have indicated that where high sodium levels in water exist, pod rot of peanut can be reduced with applications of gypsum. Although water relations in the peanut pod rot system have not been studied extensively, it is well accepted that calcium must be in an aqueous form to be absorbed by the fruit. From the data presented on nutrition as a factor in peanut pod rot (5,9,10,21,32,33,51) it appears that pod rot or pod breakdown is similar to blossom-end rot as described on other fruits, with the dissimilarity being that the peanut fruit develops beneath the soil. Peanut fruit are subterranean and they are very susceptible to colonization by microorganisms following predisposition by a nutrient imbalance or deficiency. Unquestionably, microorganisms are involved in the expression of the symptoms associated with pod rot. *Pythium myriotylum, R. solani* and *S. rolfsii* are documented peanut pathogens, which may (under undefined conditions) cause pod decay even in the absence of a nutrient predisposition (13,14,16,18,19,22,23,24,25,26,27).

Evidence proposing that nutrient imbalances are an important determinant of pod rot are abundant from the data collected by several researchers from Virginia, Georgia and Texas. Although controversy from other peanut growing states exists as to the primary cause of peanut pod rot, the data on the involvement of plant nutrition are convincingly strong. The major weaknesses of advocating a pure pathological cause of the disease complex are: first, the poor relationship between inoculum and disease response; second, the lack of good selective media to isolate colony forming propagules from soil and tissue, making population data difficult to interpret; and third, conditions which are inherent to the peanut pod rot complex are eliminated when sterile greenhouse, or sterile microplot techniques are employed to study the problem. Current recommendations for pod rot complex control in most eastern peanut growing states include the use of high rates of gypsum applied at flowering time. Nutritional imbalances in the pod development zone also are discouraged. Chemical control is not effective and may be too expensive. However, in other areas of the country chemical control measures are being recommended. In these areas, the disease usually is considered pathological in nature and only one pathogen is responsible for the majority of the disease.

Many soilborne fungi can cause decay of peanut pods. Determining the causal agent of each disease can be accomplished to a large extent by the symptoms found on stems, pods and roots. The pod rot complex is distinct in that usually no root or stem decay accompanies the pod decay. Garren first described the disease as "pod breakdown." If the disease is considered in this light the evidence for a nutritional imbalance predisposition appears to be relevant to his first descriptions of the disease.

LITERATURE CITED

1. Ashworth, L.J. Jr., and Langley, B.C. 1964. The relationship of pod damage to kernel damage by molds in Spanish peanut. Plant Dis. Rep. 48:875-878.
2. Bateman, D.F. 1964. An induced mechanism of tissue resistance to polygalacturonase in *Rhizoctonia*-infected hypocotyls of bean. Phytopathology 54:438-445.
3. Bledsoe, R.W., Comer, C.L., and Harris, H.C. 1949. Absorption of radio-active calcium by the peanut fruit. Science 109:329-330.
4. Bledsoe, R.W., and Harris, H.C. 1950. The influence of mineral deficiency on vegetative growth, flower, and fruit production and mineral composition of the peanut plant. Plant Physiol. 25:63-77.
5. Boswell, T.E., and Thames, W.H. 1976. *Pythium* pod rot control in South Texas. Proc. Am. Peanut Res. Educ. Assoc. 8:89.
6. Brady, N.C., and Colwell, W.E. 1945. Yield and quality of large-seeded type peanuts as affected by potassium and certain combinations of potassium, magnesium and calcium. J. Am. Soc. Agron. 37:429-442.
7. Brady, N.C. 1947. The effect of period of calcium supply and mobility of calcium in the plant on peanut fruit filling. Soil Sci. Soc. Am. Proc. 12:336-338.
8. Colwell, W.E., and Brady, N.C. 1945. The effect of calcium on yield and quality of large- seeded type peanuts. J. Am. Soc. Agron. 37:413- 428.

9. Csinos, A.S., Gaines, T.P., and Walker, M.E. 1984. Involvement of nutrition and fungi in the peanut pod rot complex. Plant Dis. 68:61-65.

10. Csinos, A.S., and Gaines, T.P. 1986. Peanut Pod Rot Complex: A geocarposphere nutrient imbalance. Plant Dis. 70:525-529.

11. DeKock, C.P., Hall, A., Boggie, R., and Inkson, R.H.E. 1982. The effect of water stress and form of nitrogen on the incidence of blossom-end rot in tomatoes. J. Sci. Food Agric. 33:509-515.

12. Evans, H.J., and Troxler, R.V. 1953. Relation of calcium nutrition to the incidence of blossom- end rot in tomatoes. Proc. Am. Soc. Hort. Sci. 61:346-352.

13. Frank, Z.R. 1967. Effect of irrigation procedure on *Pythium* pod rot of groundnut pods. Plant Dis. Rep. 51:414-416.

14. Frank, Z.R. 1968. *Pythium* pod rot of peanut. Phytopathology 58:542-543.

15. Frank, Z.R., and Kirkun, J. 1969. Evaluation of peanut (*Arachis hypogaea*) varieties for Verticillium wilt resistance. Plant Dis. Rep. 53:744-746.

16. Frank, Z.R. 1972. *Pythium myriotylum* and *Fusarium solani* as cofactors in a pod-rot complex of peanut. Phytopathology 62:1331-1334.

17. Frezzi, M.J. 1956. Especies de *Pythium* fitopatogenas identificadas en la Republica Argentina. Rev. Invest. Agric., Buenos Aires 10:113-241.

18. Garcia, R., and Mitchell, D.J. 1975. Synergistic interactions of *Pythium myriotylum* with *Fusarium solani* and *Meloidogyne arenaria* in pod rot of peanut. Phytopathology 65:832-833.

19. Garcia, R., and Mitchell, D.J. 1975. Interactions of *Pythium myriotylum* with several fungi in peanut pod rot. Phytopathology 65:1375- 1381.

20. Garren, K.H. 1964. Isolation procedures influence the apparent make-up of the terrestrial microflora of peanut pods. Plant Dis. Rep. 48:344-348.

21. Garren, K.H. 1964. Land plaster and soil rot of peanut pods in Virginia. Plant Dis. Rep. 48:349-352.

22. Garren, K.H. 1964. Recent developments in research on peanut pod rot. Proc. 3rd Nat'l. Peanut Res. Conf. Auburn, AL, pp. 20-27.

23. Garren, K.H. 1966. Controlling pod rot. The Peanut Farmer 2(9):16. (Raleigh, NC) August.

24. Garren, K.H. 1966. Peanut (groundnut) microfloras and pathogenesis in peanut pod rot. Phytopathol. Z. 55:359-367.

25. Garren, K.H. 1967. Relation of several pathogenic organisms, and the competition of *Trichoderma viride* to peanut pod breakdown. Plant Dis. Rep. 51:601-605.
26. Garren, K.H. 1970. *Rhizoctonia* solani versus *Pythium myriotylum* as pathogens of peanut pod breakdown. Plant Dis. Rep. 54:840-843.
27. Garren, K.H. 1970. Antagonisms between indigenous *Pythium myriotylum* and introduced *Rhizoctonia solani* and peanut pod breakdown. Phytopathology 60:1292 (Abstr.).
28. Geraldson, C.M. 1957. Control of blossom-end rot of tomatoes. Proc. Am. Soc. Hort. Sci. 69:309-317.
29. Geraldson, C.M. 1957. Factors affecting calcium nutrition of celery, tomato and pepper. Proc. Soil Sci. Soc. Am. 21:621-625.
30. Gerard, C.J., and Hipp, B.W. 1968. Blossom- end rot of "Chico" and "Chico Grande" tomatoes. Proc. Am. Soc. Hort. Sci. 93:521-531.
31. Hallock, D.L. 1962. Effect of time and rate of fertilizer application on yield and seed-size of jumbo runner peanuts. Agron. J. 54:428-430.
32. Hallock, D.L., and Garren, K.H. 1968. Pod breakdown, yield, and grade of Virginia type peanuts as affected by Ca, Mg, and K sulfates. Agron. J. 60:253-257.
33. Higgins, B.B. 1954. In Growing Peanuts. J.H. Beattie, F.W. Poos, and B.B. Higgins, eds. U. S. Dept. Agric. Farm Bull. p. 2063.
34. Jacobson, L., Hannapel, R.J., Moore, D.P., and Schaedle, M. 1961. Influence of calcium on the selectivity of the ion absorption process. Plant Physiol. 36:58-61.
35. Kranz, G., and Pucci, E. 1963. Studies on soil- borne rots of groundnuts (*Arachis hypogaea*). Phytopathol. Z. 47:101-112.
36. McColloch, L.P., Cook, H.T., and Wright, W.R. 1968. Market disease of tomatoes, peppers and eggplants. Agric. Res. Serv. U. S. Dep. Agric., Agric. Handb. 28:10-11.
37. Middleton, G.F., Colwell, W.E., Brady, N.C., and Schultz, E.F. Jr. 1945. The behavior of four varieties of peanuts as affected by calcium and potassium variables. J. Am. Soc. Agron. 37:443-457.
38. Moore, L.D., and Wills, W.H. 1974. The influence of calcium on the susceptibility of peanut pods to *Pythium myriotylum* and *Rhizoctonia solani*. Peanut Sci. 1:18-20.
39. Porter, D.M., and Smith, J.C. 1974. Fungal colonization of peanut fruit as related to southern corn rootworm injury. Phytopathology 64:249-251.

40. Porter, D.M., Smith, D.H., and Rodriguez-Kabana, R. 1982. Peanut plant diseases: soilborne diseases. Pages 348-378 in: Peanut Science and Technology, H.E. Pattee and C.T. Young, eds., Am. Peanut Res. Educ. Soc., Yoakum, TX.

41. Raleigh, S.M., and Chucka, J.A. 1944. Effect of nutrient ratio and concentration on growth and composition of tomato plants and on the occurrence of blossom-end rot of fruit. Plant Physiol. 19:671-678.

42. Reed, J.F. and N.C. Brady, N.C. 1948. Time and method of supplying calcium as factors affecting production of peanuts. J. Am. Soc. Agron. 40:980- 996.

43. Robertson, W.K., Lundy, H.W., and Thompson, L.G. 1965. Peanut responses to calcium sources and micronutrients. Soil Crop Sci. Soc. Fla. Proc. 25:335-343.

44. Shew, H.D., and Beute, M.K. 1979. Evidence for the involvement of soilborne mites in *Pythium* pod rot of peanut. Phytopathology 69:204-207.

45. Skelton, B.J., and G.M. Shear. 1971. Calcium translocation in the peanut (*Arachis hypogaea* L.). Agron. J. 63:409-412.

46. Sullivan, G.A., Jones, G.L., and Monroe, R.P. 1974. Effect of dolomitic limestone, gypsum, and potassium on yield and seed quality of peanuts. Peanut Sci. 1:73-77.

47. Walker, M.E. 1975. Calcium requirements for peanuts. Commun. Soil. Sci. Plant Anal. 6(3):299- 313.

48. Walker, M.E., Keisling, T.C., and Drexler, J.S. 1976. Responses of three peanut cultivars to gypsum. Agron. J. 68:527-528.

49. Walker, M.E., and Keisling, T.C. 1978. Response of five cultivars to gypsum fertilization on soils varying in calcium content. Peanut Sci. 5:57-60.

50. Walker, M.E., Flowers, R.A., Henning, R.J., Keisling, T.C., and Mullinix, B.G. 1979. Response of Early Bunch peanuts to calcium and potassium fertilization. Peanut Sci. 6:119-123.

51. Walker, M.E., and Csinos, A.S. 1980. Effect of gypsum on yield, grade, and incidence of pod rot in five peanut cultivars. Peanut Sci. 7:109- 113.

52. Yang, C.Y.D., and Mitchell, J.E. 1965. Cation effect on reproduction of *Pythium* spp. Phytopathology 55:1127-1131.

SOME EFFECTS OF MINERAL NUTRITION ON AFLATOXIN CONTAMINATION OF CORN AND PEANUTS

David M. Wilson, Milton E. Walker
and Gary J. Gascho
Departments of Plant Pathology & Agronomy
University of Georgia
Coastal Plain Experiment Station
Tifton, Georgia 31793

The influence of mineral nutrition on preharvest aflatoxin contamination has been studied in only a few crops. Both N nutrition of corn and Ca nutrition of peanuts have been shown to affect preharvest aflatoxin contamination in certain locations and soil types. The purpose of this chapter is to review the literature on the effects of mineral nutrition on preharvest aflatoxin contamination and to present previously unpublished experiments on peanuts and corn.

AFLATOXIN CONTAMINATION OF CORN

Aflatoxin contamination of corn as a result of infection by the *Aspergillus flavus* group[1] in the field was recognized and studied by the Quaker Oats Company in 1971, 1972, and 1973 because contamination of feed could not be attributed to poor handling and storage practices. Anderson, Nehring and Wichser (1) published the Quaker Oats study in 1975 and included observations on insect damage, location, plant populations and fertilization. They observed more aflatoxin contamination in corn from the warmer, more humid corn growing regions of the United States. Conditions leading to plant stress, including dense populations of plants and reduced fertilization, appeared to result in more aflatoxin contamination in the field (1).

[1]*A. flavus* group refers to either *A. flavus* Link or *A. parasiticus* Speare, or both. For convenience *A. flavus* generally means the *A. flavus* group in this paper.

Jones and Duncan (10) measured the effects of N fertilization, planting date and harvest date on aflatoxin contamination of inoculated field corn. Less aflatoxin B1 was detected in treatments receiving 145 kg N ha^{-1} than in treatments receiving 11 kg N ha^{-1}. Corn from low N plots had 2.4 times more aflatoxin B1 than corn from high N plots when averaged over cultivar, planting date and *A. flavus* isolate. Jones and Duncan (10) suggested that inadequate N fertilization alters the nutritional status of preharvest corn, making it a good substrate for aflatoxin production. In Georgia, during the drought year 1977, Wilson and Keisling (unpublished) observed more aflatoxin in low N plots from both irrigated and non-irrigated corn (Table 1). Younis *et al.* (19) demonstrated that drought stress alters the uptake and translocation of N in corn.

Table 1. Aflatoxin content of corn grain in 1977 as affected by rate and type of nitrogen fertilizer.*

Nitrogen Source	kg ha^{-1} Nitrogen	ng gm^{-1} Aflatoxin
None	0	637 a**
Ammonium nitrate	112	20 b
Ammonium nitrate	224	3 b
Ammonium nitrate	336	10 b
Sulfur coated urea	336	12 b

*   Values from irrigated and non-irrigated plots, irrigation in 1977 had no significant effect on aflatoxin content.

** Numbers followed by the same letter are not significantly different at P = 0.05 by the Least Significant Difference Test.

Fortnum and Manwiller (6) and Payne *et al.* (14) studied the effects of irrigation on aflatoxin production in corn. Irrigation suppressed aflatoxin production, and Fortnum and Manwiller (6) suggested that differences between years were due to strong environmental effects on aflatoxin contamination. Payne *et al.* (14) reported that irrigation and subsoiling both reduced aflatoxin contamination. Jones *et al.* (11) found that irrigation resulted in fewer airborne conidia and infected kernels. There are high inoculum levels of *A. flavus* and severe environmental stresses in Georgia's deep sandy soils. These factors may help explain the differences among the Georgia, North Carolina and South Carolina observations on the effects of irrigation. In some years, it may be very difficult to apply enough water to overcome water stress in corn in Georgia grown on deep sand.

Aflatoxin contamination of corn in studies in Tifton, Georgia, has ranged from 0 to 637 ppb total aflatoxins. None of the corn grown under center pivot irrigation with recommended fertility contained more than 50 ppb of total aflatoxins, whereas some of the corn grown under other irrigation regimes contained over 500 ppb of total aflatoxins. It appears that center pivot irrigation helps minimize aflatoxin, but the reasons for this are not immediately apparent. Other studies need to be conducted to compare the effects of center pivot irrigation with other forms of irrigation to further define the effects of irrigation and environment. Variations observed in aflatoxin contamination in field studies make comparisons of treatments difficult, and the differences seen may be due to interactions of biological factors with temperature and moisture.

Tillage practices, nematode control and fertility other than N have had no significant effects on aflatoxin contamination. However, data from low N or late N treatments indicate that N may influence aflatoxin contamination at harvest.

Aflatoxin concentrations have been determined for corn in several plant nutrition experiments at Tifton. Since 1980, all experiments were irrigated. With normal fertilization levels, adequate irrigation and rapid drying of harvested corn, levels of aflatoxin have been low. However, in 1981 and 1986 experiments on Bonifay sand (loamy, siliceous, thermic, gross-arenic Plinthic Paleudult) the irrigation was marginal, due to dry growing seasons, and extreme rates of nutrients were applied. During those two years, aflatoxins varied significantly with nutrients applied.

In 1981 Pioneer 3369A corn was planted in 0.61 meter rows on March 9. One-half of the plots were thinned from 89 to 79 thousand plants ha^{-1} at the V6 stage. All nutrient applications, except N, and all cultural practices were uniform and as recommended for irrigated corn by the Georgia Extension Service. The total irrigation for the crop was 38 cm. Corn was hand- harvested in late July and shelled. The grain was dried to <15% moisture and yields were calculated to 15.5%.

Since aflatoxin measurements always have high variability, large differences in means were required for significance. Significant differences were found at the higher population (89 thousand plants ha^{-1}). (Table 2). The late fertigation at 10 weeks of age resulted in high N concentrations late in the season and high aflatoxin. Check plots in this experiment yielded only 4.7 MT ha^{-1} and yield in the fertigated plots responded in a quadratic fashion showing maximum yield of 11.4 MT ha^{-1} at 322 kg N ha^1.

In 1986, the experiment was designed with the objective of making in-crop corrections of nutrient deficiencies or imbalances via

139

fertigation based on foliar analyses at the 10-leaf stage. The experiment was a split-plot design and was conducted with four initial fertility regimes that approximately corresponded to 0.5, 0.75, 1.0 and 1.5 times that recommended by Georgia Extension Service as subplots and six supplemental fertigation treatments. Pioneer 3165 corn was planted in double rows that were 18 cm apart on 90 cm centers on March 25. The initial P levels were established with the 10-34-0 starter placed 5 cm to the side and 5 cm below the seed. Higher P levels were established by hand application of 0-46-0. One-half the K was applied as 0-0-62 with the herbicide (co-chemigation) following planting and one-half by fertigation at the V8 stage. Nitrogen differentials were applied at the V6 stage by dribbling 32-0-0 next to the row. The remainder of the N was fertigated according to a standard fertigation schedule. The supplemental applications were applied with an irrigation simulator at the V12 stage. Twenty-one cm of irrigation were supplied during the season. The plots were combine harvested on August 12, and the grain handled as in 1981.

Table 2. Effect of the rate and method of nitrogen application on total aflatoxin content in corn grain grown at two populations in Bonifay sand in 1981.

Nitrogen treatment		Plants ha^{-1} x 1000 Plant population	
		79	89
Method	Nitrogen	-----Aflatoxin----	
	kg ha^{-1}	ng gm^{-1}	
Check[v]	87	55 a	32 b[z]
Fertigate[w]	168	10 a	6 b
Fertigate	224	25 a	0 b
Fertigate	280	21 a	2 b
Fertigate	336	3 a	2 b
Late fertigate[x]	224	19 a	90 a
Late conventional[y]	224	13 a	51 ab
Late conventional	336	45 a	41 ab

[v] Check received all 87 kg N at planting.
[w] Fertigated plots received 87 kg N at planting and the remaining split in 3 equal applications in weeks 4, 6, and 8 following planting via the set irrigation system.
[x] Late fertigated plots received only one fertigation at 10 weeks plus the original 87 kg at planting.
[y] Late conventional plots received one application of ammonium nitrate at 10 weeks plus the original 87 kg at planting.
[z] Values within a group not followed by a common letter are significantly different at the 0.05 level by Waller-Duncan K ratio test.

Aflatoxin content of the corn was related to the initial NPK fertility treatment (Table 3). At a fertility level higher than recommended, aflatoxin concentration was significantly higher than at recommended (or lower) fertility levels. Yield was highest for the recommended levels of N, P and K. Thus, fertility applied at an excessive level appeared to increase aflatoxin in the grain.

Table 3. Effects of initial fertility level on aflatoxin in corn grown on Bonifay sand in 1986.*

Initial applied fertility			Total
N	P	K	aflatoxins
	kg ha^{-1}		ng gm^{-1}
140	25	60	65 b**
211	37	91	56 b
280	49	121	42 b
364	74	81	148 a

* No significant differences for fertility level - supplemental fertility interaction at the 0.10 level by the F test.

** Values within a group and a column followed by a common letter are not significantly different by the Least Significant Difference Test at the 0.10 level.

Table 4. Supplemental fertilizer effects on aflatoxin in corn grown on Bonifay sand in 1986.*

Supplemental fertilizer applied				Total
N	Ca	Mg	S	aflatoxins
	kg ha^{-1}			ng gm^{-1}
0	0	0	0	49 b**
45	0	0	0	66 ab
90	0	0	0	85 ab
45	0	11	0	86 ab
45	11	0	0	50 b
45	0	0	6	130 a

* No significant differences for fertility level - supplemental fertility interaction at the 0.10 level by the F test.

** Values within a group and a column followed by a common letter are not significantly different by the Least Significant Difference Test at the 0.10 level.

Application of S as ammonium-thiosulfate at the 12-leaf stage also increased aflatoxin in this trial (Table 4). There was no interaction between initial fertility level and supplemental fertilization.

Thus, the Tifton experiments showed significant differences in aflatoxin concentration due to nutrition. The data suggest that aflatoxin concentrations are increased when plants are stressed due to inadequate irrigation, high population and/or extreme N deficiency.

Furthermore, the late applications of large amounts of N or excessive fertilization with N, P and K also resulted in elevated aflatoxin concentrations.

AFLATOXIN CONTAMINATION OF PEANUTS

More is known about the influence of environmental stress on aflatoxin contamination in peanuts than in corn. Detailed studies on the effects of temperature and soil moisture conditions on colonization of kernels and aflatoxin contamination of peanuts have been carried out in plots at Dawson, Georgia (3,8,15). Little aflatoxin contamination occurred in peanuts from plots without drought stress in studies in Dawson or Tifton, Georgia (17,19), but aflatoxin contamination was high in drought plots maintained between 25 and 32 C (9). Aflatoxin and colonization by the *A. flavus* group were maximal in plots with a mean geocarposphere temperature of 30.5 C. Edible grades of kernels from irrigated plots contained no aflatoxins, but any damage to pods or kernels increased both *A. flavus* colonization and aflatoxin contamination. Blankenship *et al.* (2) found that the greatest aflatoxin concentration corresponded to the highest incidence of *A. flavus* in drought stressed sound mature kernels (SMK).

Sanders *et al.* (15) measured pod and stem temperatures in the moisture and temperature controlled plots at Dawson. They demonstrated that peanut stems in drought treatments reached a maximum of about 40 C. For 6-7 hours of each day these stems were as much as 10 C warmer than stems of plants in irrigated plots. They also found that as pod temperatures approached 35 C, the proportion of kernels colonized by *A. flavus* and aflatoxin contamination increased.

Gaines and Hammons (7) compared the mineral composition of peanut seed from four cultivars grown at six locations in the southern United States. The seeds were analyzed for ten nutrients: P, K, Ca, Mg, S, Zn, Mn, Cu, Fe and B. Significant differences were found in the amounts of nine nutrients among locations and eight nutrients among cultivars. The only positive correlations found for locations were between seed Ca and total precipitation and between seed Mg and soil Mg.

Because severe drought conditions can result in Ca and B deficiency symptoms in peanuts, Wilson and Walker (18) determined the effect of Ca, in the form of gypsum, on aflatoxin contamination of peanuts. They found that gypsum ($CaSO_4.2H_2O$) applications equivalent to Ca rates of 112, 224, and 332 kg ha^{-1} significantly reduced aflatoxin contamination when compared to peanuts grown without applied Ca. Unfortunately, similar comparisons could not be made in other years because no aflatoxin contamination occurred. Mixon *et al.* (13) found no aflatoxin in gypsum treated plots, but aflatoxin was occasionally found in peanuts from the non-gypsum treatments resulting in a highly significant treatment and genotype interaction. Gypsum applications of 673 and 1345 kg ha^{-1} reduced the percentage of seed colonized by the *A. flavus* group and enhanced control of seed colonization in plots treated with *Trichoderma harzianum*.

Davidson *et al.* (4,5) studied the effects of row spacing, row orientation and gypsum on quality of peanut seed from non-irrigated fields. They found that the application of gypsum (224 kg ha^{-1} of Ca) at bloom, north-south row orientation and close row spacing provided certain benefits (4). Gypsum applied at bloom reduced kernel damage, foreign material, and improved quality and germination.

Close spaced rows provided a larger tap root crop, cooler soil temperatures and slightly higher germination percentages than wider rows. North-south row orientations provided cooler soil temperatures, higher yields and greater germination percentages than east- west row orientations. Application of gypsum at bloom increased germination and reduced aflatoxin contamination by a factor of two. It is apparent from the results presented by Davidson *et al.* (5) that applications of Ca, as gypsum at bloom, may lower but not eliminate aflatoxin contamination in non- irrigated peanuts.

The effects of gypsum applications on preharvest aflatoxin contamination of peanuts have been variable in experiments conducted at Tifton, Georgia. In years with aflatoxin contamination, applications of gypsum at early bloom have always lowered aflatoxin amounts, but in many years no aflatoxins were detected for any treatment. Because of the variable nature of field aflatoxin contamination, Wilson and Walker designed experiments to test the effects of irrigation, gypsum rates and inoculation with *A. parasiticus* on *A. flavus* group populations and aflatoxin contamination at harvest. Field experiments were conducted in 1984 and 1985 in Tifton, Georgia, on a Lakeland sand (Thermic, coated Typic Quartzipsamments). The experiments used a split-split plot design. Irrigation and non-irrigation were the whole plots; split plots consisted of various rates (0, 560, 1120, 1680 kg ha^{-1}) of gypsum corresponding and to 0, 112, 224, and

336 kg ha^{-1} of added Ca The split-split plots were inoculated or not with *A. parasiticus* NRRL 2999. The plots were two rows spaced 0.91 meters apart and 6.09 m long. The preplant fertilization was 560 kg ha^{-1} of 5-10-15 fertilizer broadcast and incorporated before planting. Florunner peanuts were planted on May 19, 1984, and NC-7 peanuts were planted on May 22, 1985. All recommended practices for peanut production were carried out during the growing season. Calcium was applied as gypsum in a 30 cm band centered over the row at the early bloom stage (55 days after planting in 1984 and 64 days after planting in 1985). Soil water pressure on the irrigated plots was maintained between -10 to -60 KPa before pegging and from -10 to -30 KPa after pegging. Soil samples were taken three times from 0-15 cm depth and analyzed for pH, P, K, Ca and Mg.

An *A. parasiticus* spore suspension was applied on July 12, 1984, and July 24, 1985, on inoculated plots using a watering can centered over the row. Soil samples were taken four times each year to determine *A. flavus* group soil populations.

Peanuts were dug October 11, 1984 and September 23, 1985, and allowed to dry in a window. Combined peanuts were dried at 43 C until they reached 8.5% moisture. A 500 g sample was used for grade determinations based on Federal State Inspection Service guidelines for determining percentage of sound mature kernels. The value per hectare was based on yield, quality factors and support price.

Hulls and kernels from each plot were analyzed for P, K, Ca and Mg. Hulls and kernels were plated and incubated for 7 days at 30 C on malt salt agar after surface disinfestation with 0.5% aqueous NaOC1. Aflatoxin amounts in 50 g of kernels from each plot were determined using the HPLC method developed by Thean *et al.* (16).

Table 5. Effect of Ca (gypsum) on peanut yield and sound mature kernels in Florunners in 1984 and NC-7 in 1985.*

Calcium	Yield		SMK (%)	
	1984	1985	1984	1985
kg ha^{-1}	kg ha^{-1}			
0	4965 b**	1635 c	73.2 b	64.3 b
112	5990 a	2819 b	75.9 a	70.8 a
224	5972 a	3212 a	76.9 a	72.4 a
336	6238 a	2987 ab	76.8 a	72.7 a

*   Calcium applied as gypsum at early bloom.
** Means within a column with different letters are significantly different by the Least Significant Difference test, (P = 0.05).

Application of gypsum in the 1984 and 1985 tests increased yield, value and SMK of peanuts grown on Lakeland sand (Table 5) and significantly reduced damage in 1985 from 3.75% with no Ca added to 2.06% with 336 kg ha^{-1} Ca. There was no effect of irrigation on yield, value and SMK. *A. parasiticus* inoculation significantly decreased Florunner yield from 5895 kg ha^{-1} to 5688 kg ha^{-1} in 1984, but no significant yield decrease was seen in 1985 with NC-7.

Gypsum treated plots had significantly higher Ca in the soil than the untreated plots. However, differences among rates were not significant. Potassium and Mg contents decreased in soil from plots receiving Ca. This reduction was expected due to the interactions between Ca and other cations.

Inoculation of peanut plants by sprinkling a spore suspension of *A. parasiticus* over the plants at early bloom increased the populations of the *A. flavus* group in the soil in both 1984 and 1985. The differences persisted throughout the season in 1985, but differences were not apparent with the September 5 and September 27 collection dates in 1984 (Table 6). Irrigation and Ca treatments did not have significant effects on the soil populations of the *A. flavus* group.

Irrigation and gypsum applications had different effects on P, K, Ca and Mg concentrations in hulls and kernels of NC-7 peanuts (Tables 7 & 8). Irrigation significantly increased P in kernels and hulls, significantly decreased K in hulls, significantly increased Ca in hulls and kernels, and significantly increased Mg in kernels. Gypsum applications decreased P and Mg in hulls and K in kernels and significantly increased Ca content of hulls and kernels.

Table 6. Effect of *Aspergillus parasiticus* (NRRL 2999) inoculation on *A. flavus* group soil populations.

	Sampling Date 1984				Sampling Date 1985			
	7/24	8/15	9/5	9/27	7/31	8/14	9/4	9/25
Not Inoculated[y]	88 b[z]	62 b	132	84	50 b	30 b	46 b	46 b
Inoculated	138 a	132 a	146	98	166 a	108 a	112 a	120 a

[y] Plants inoculated by sprinkling plants with a conidial suspension at early bloom.

[z] Means within a column with different letters are significantly different by the Least Significant Difference Test, (P = 0.05).

Irrigation and gypsum treatments both affected seed and hull colonization by the *A. flavus* group. Table 9 summarizes the data obtained from kernels with NC-7 peanuts in 1985. The recovery of fungi from hulls was similar.

Irrigation significantly reduced the numbers of kernels and hulls from which members of the *A. flavus* group and *A. niger* were recovered. Recovery of all other fungi did not seem to be affected by irrigation although no attempt was made to distinguish genera.

Gypsum applications (Table 10) affected peanut seed colonization of Florunner peanuts in 1984. Application of Ca reduced percent colonization of seed (kernels) from 7.38% to 4.06% with 112 and 336 kg ha^{-1} of Ca. The highest Ca application significantly decreased the percent of other fungi recovered from the seed.

INFLUENCE OF MINERAL NUTRITION

Environmental conditions in combination with mineral nutrition affect aflatoxin contamination of both corn and peanuts. The nutrient status of the plant influences both infection and aflatoxin production by *A. flavus*.

Table 7. Influence of irrigation and calcium (gypsum) on nutrient concentration of hulls of NC-7 peanuts.

Treatment		% Nutrient content of peanut hulls			
Rate[y]					
Irrigation	Calcium	P	K	Ca	Mg
	kg ha^{-1}				
1	0	0.071	0.486	0.106	0.075
1	112	0.050	0.483	0.126	0.060
1	224	0.038	0.466	0.162	0.051
1	336	0.047	0.462	0.187	0.052
Avg		0.051	0.474	0.145	0.059
2	0	0.078	0.450	0.128	0.075
2	1120	0.050	0.410	0.131	0.056
2	224	0.048	0.405	0.163	0.052
2	336	0.054	0.444	0.202	0.057
Avg		0.057	0.427	0.156	0.060
Significance[z]					
Irrigation		*	**	----	----
Calcium		**	—	**	**

[y] 1 = No irrigation and 2 = Irrigated when needed as indicated by soil tensiometer.

[z] * and ** indicate significance of the F test at 0.05 and 0.01, respectively.

Table 8. Influence of irrigation and calcium (gypsum) on nutrient concentration of kernels of NC-7 peanuts.

Treatment			% Nutrient content in peanut hulls		
Rate[y]					
Irrigation	Calcium	P	K	Ca	Mg
	kg ha⁻¹				
1	0	0.353	0.727	0.036	0.151
1	112	0.343	0.743	0.031	0.151
1	224	0.330	0.707	0.038	0.151
1	336	0.032	0.696	0.040	0.151
Avg.		0.339	0.718	0.036	0.151
2	0	0.364	0.706	0.036	0.163
2	112	0.360	0.703	0.038	0.155
2	224	0.346	0.672	0.042	0.155
2	336	0.361	0.674	0.048	0.155
Avg.		0.357	0.688	0.041	0.157
Significance[z]					
Irrigation		*	-----	**	**
Calcium		----	**	**	----

[y] 1 = No irrigation and 2 = Irrigated when needed as indicated by soil tensiometer.

[z] * and ** indicate significance of the F test at 0.05 and 0.01, respectively.

Table 9. Effect of irrigation on percent fungi recovered from NC-7 peanut kernels and hulls.

	% Kernels with		
Irrigation*	A. flavus	A. niger	Other fungi
None	19.5 a**	14.0 a	95.0 a
As needed	7.8 b	5.1 b	97.5 a

* Irrigation was applied as needed according to soil tensiometer. Peanut pods were surface disinfested with 0.05% NaOCl for 5 minutes, then carefully shelled and the kernels and hulls were plated on malt salt agar.

** Mean values followed by different letters are significantly different, (P = 0.05).

However, the members of the *A. flavus* group that infect corn and peanuts are not specialized pathogens and plant stresses caused by environmental or biotic factors contribute to aflatoxin production. These stresses can overcome the beneficial effects of mineral nutrition in the warm humid regions of the United States. Lillehoj (12) reviewed the effects of several of these environmental and cultural factors on aflatoxin contamination of corn.

Adequate N nutrition of corn was shown to positively affect aflatoxin contamination in North Carolina (10) and sometimes in Georgia. There are at least two reasons for the differences in location. First, North Carolina has lower temperatures and probably lower *A. flavus* group populations during the growing season than Georgia; and N may be more effective in lowering infection of corn where the inoculum density was not overwhelming. Second, the studies in Georgia were in deep sandy soils where temperature, moisture and insect stresses probably cannot be avoided. This may explain Georgia's inability to show a consistent relationship between irrigation and aflatoxin production in corn; whereas, in North Carolina and South Carolina, irrigation consistently lowered aflatoxin concentrations (6,14). Nitrogen uptake was influenced by water availability, and drought conditions may have limited nitrogen uptake and translocation and perhaps minimized the positive influence of N in corn.

Table 10. Effect of calcium on colonization of Florunner peanut seed by fungi.*

Calcium	% Seed with *A. flavus* group	% Seed with other fungi
kg ha^{-1}		
0	7.38 a**	42.5 a
112	4.06 ab	34.1 ab
224	3.25 b	34.7 ab
336	4.06 ab	29.1 b

* Calcium applied as gypsum at early bloom in 1984 (48 DAP).
** Means within a column followed by different letters are not significantly different by the Least Significant Difference Test, (P = 0.05).

Calcium availability in peanuts can affect both *A. flavus* group interactions and aflatoxin contamination of peanut kernels. Studies in Georgia showed that Ca applications in certain soils had a beneficial effect on lowering aflatoxin contamination. This phenomena was most easily observed in Ca deficient sandy soils, but even then a consistent relationship was difficult to demonstrate because aflatoxin contamination occurred in the field only about one year out of five. It was difficult to understand the Ca relationship when perhaps only 3 of 15 tests contained aflatoxin in any treatment.

Davidson *et al.* (5) published data from a single farm in crop year 1981 illustrating the relationship between Ca content of peanut kernels, germination and aflatoxin content. James I. Davidson (personal communication) feels that the relationships with Ca content above 0.04% will have to be revised with more data, but that the general trends will remain. Thus, with increasing Ca content of the seed, there should be less aflatoxin contamination, less overall damage and higher germination. Other important relationships between peanuts and Ca uptake include water availability and soil temperature. Calcium deficiencies can be induced by drought and high soil temperatures and these factors also influence the infectivity and populations of the *A. flavus* group.

In summary, N nutrition, in combination with other production practices, has a positive influence in reducing *A. flavus* group infections and aflatoxin contamination in corn when other environmental or biological stresses are not extreme. In peanuts, the same can be said for Ca nutrition in certain soils. In sandy, Ca deficient soils, applications of Ca increase yield and quality and decrease damage in harvested peanuts. Both Ca applications and irrigation practices influence hull and kernel infection by the *A. flavus* group that are sometimes accompanied by peanuts with high concentrations of aflatoxins. Thus, mineral nutrition is only one of several production practices that may be used to help control *A. flavus* infection and aflatoxin contamination of crops.

LITERATURE CITED

1. Anderson, H.W., Nehring, E.W., and Wichser, W.R. 1975. Aflatoxin contamination of corn in the field. J. Agric. Food Chem. 23:775-782.

2. Blankenship, P.D., Cole, R.J., Sanders, T.H., and Hill, R.A. 1984. Effect of geocarposphere temperature on pre-harvest colonization of drought- stressed peanuts by *Aspergillus flavus* and subsequent aflatoxin contamination. Mycopathologia 85:69-74.

3. Cole, R.J., Hill, R.A., Blankenship, P.D., Sanders, T.H., and Garren, K.H. 1982. Influence of irrigation and drought stress on invasion by *Aspergillus flavus* of corn kernels and peanut pods. Dev. Ind. Microbiol. 23:229-236.

4. Davidson, J.I., Blankenship, P.D., Sanders, T.H., Cole, R.J., Hill, R.A., Henning, R.J., and Guerke, W.R. 1983. Effect of row spacing, row orientation, and calcium (gypsum) on seed quality. Proc. GCIA Ann. Meet. 4:73-79.

5. Davidson, J.I., Blankenship, P.D., Sanders, T.H., Cole, R.J., Hill, R.A., Henning, R.J., and Guerke, W.R. 1983. Effect of row spacing, row orientation and gypsum on the production and quality of non-irrigated Florunner peanuts. Proc. Am. Peanut Res. Ed. Soc. 15:46-51.

6. Fortnum, B.A., and Manwiller, A. 1985. Effects of irrigation and kernel injury on aflatoxin B1 production in selected maize hybrids. Plant Dis. 69:262-265.

7. Gaines, T.P., and Hammons, R.O. 1981. Mineral composition of peanut seed as influenced by cultivar and location. Peanut Sci. 8:16-20.

8. Hill, R.A., Blankenship, P.D., Cole, R.J., and Sanders, T.H. 1983. Effects of soil moisture and temperature on preharvest invasion of peanuts by the *Aspergillus flavus* group and subsequent aflatoxin development. Appl. Environ. Microbiol. 45:628-633.

9. Hill, R.A., Wilson, D.M., McMillian, W.W., Widstrom, N.W., Cole, R.J., Sanders, T.H., and Blankenship, P.D. 1985. Ecology of the *Aspergillus flavus* group and aflatoxin formation in maize and groundnut. Pages 79-95 in: Trichothecenes and Other Mycotoxins. J. Lacey, Ed., John Wiley & Sons, London.

10. Jones, R.K., and Duncan, H.E. 1981. Effect of nitrogen fertilizer, planting date, and harvest date on aflatoxin production in corn inoculated with *Aspergillus flavus*. Plant Dis. 65:741-744.

11. Jones, R.K., Duncan, H.E., and Hamilton, P.B. 1981. Planting date, harvest date, and irrigation effects on infection and aflatoxin production by *Aspergillus flavus* in field corn. Phytopathology 71:810-816.

12. Lillehoj, E.B. 1983. Effect of environmental and cultural factors on aflatoxin contamination of developing corn kernels. Pages 27-34 in: Aflatoxin and *Aspergillus flavus* in Corn. U.L. Diener, R.L. Asquith, and J.W. Dickens, Eds. So. Coop. Ser. Bull. 279, Auburn Univ., Alabama.

13. Mixon, A.C., Bell, D.K., and Wilson, D.M. 1984. Effect of chemical and biological agents on the incidence of *Aspergillus flavus* and aflatoxin contamination of peanut seed.

Phytopathology 74:1440-1444.
14. Payne, G.A., Cassel, D.K., and Adkins, C.R. 1986. Reduction of aflatoxin contamination in corn by irrigation and tillage. Phytopathology 76:679-684.
15. Sanders, T.H., Blankenship, P.D., Cole, R.J., and Hill, R.A. 1984. Effect of soil temperature and drought on peanut pod and stem temperatures relative to *Aspergillus flavus* invasion and aflatoxin contamination. Mycopathologia 86:51-54.
16. Thean, J.E., Lorenz, D.R., Wilson, D.M., Rogers, K., and Gueldner, R.C. 1980. Extraction, cleanup and quantitative determination of aflatoxins in corn. J. Assoc. Off. Anal. Chem. 63:631-633.
17. Wilson, D.M., and Stansell, J.R. 1983. Effects of irrigation regimes on aflatoxin contamination in peanut pods. Peanut Sci. 10:46-54.
18. Wilson, D.M., and Walker, M.E. 1981. Calcium potential aflatoxin foe. Southeastern Peanut Farmer 19:6B.
19. Younis, M.A., Pauli, A.W., Mitchell, H.L., and Strickler, S.C. 1965. Temperature and its interaction with light and moisture in nitrogen metabolism of corn (*Zea mays* L.) seedlings. Crop Sci. 5:321-326.

MANAGEMENT OF COMMON SCAB OF POTATO WITH PLANT NUTRIENTS

Anthony P. Keinath and Rosemary Loria
Department of Plant Pathology
Cornell University, Ithaca, NY 14853-5908

This chapter addresses attempts at using plant nutrients for control of common scab of white potato (*Solanum tuberosum* L.) tubers, caused by the actinomycete *Streptomyces scabies* (Thaxter) Waksman et Henrici. The reader also is referred to a review by Wenzl (61) on control of several tuber diseases with cultural practices, including effects of plant nutrients on control of common scab.

Common scab is characterized by corky lesions on the tuber surface. Though other *Streptomyces* spp. have been reported to incite symptoms which are similar to those caused by *S. scabies* (37,56), this review is limited to research reported on *S. scabies*, as described by Corbaz (13). However, the taxonomy of this genus is complex (32) and characteristics necessary for verification of the species designation of the pathogenic isolates used in the studies reviewed here are generally not available in the literature.

Symptoms of potato scab are variable: lesions are usually discrete but may cover most of the tuber surface and can be superficial, slightly raised or pitted. Many different rating systems have been used to evaluate the incidence and severity of this disease. Disease incidence on tubers, generally expressed as percent of tubers with scab symptoms, is the simplest and one of the most commonly used rating systems. However, some researchers have used indices which incorporate lesion type (superficial, raised, slightly pitted or deeply pitted) (12) or lesion surface area (16,44,58). Inconsistencies in disease rating systems complicate interpretation of the data and often make it difficult to compare results from different studies.

Information about plant nutrients in relation to potato scab is generally limited to the effects of macronutrients and micronutrients on disease, nutrient concentrations in soil and, infrequently, nutrient concentrations in potato tubers. Because of the difficulty in quantitatively recovering and identifying *S. scabies* from natural soil, few data are available on the response of pathogen populations to added nutrients. Further, information available from the literature is usually not adequate to determine the mechanism of control of *S. scabies*, when control is demonstrated.

Much of the literature on potato scab control deals with the application of nutrients to the soil which also affect soil pH. Disease incidence and severity generally increase as soil pH increases from about 5.0 to 8.0, though scab occurs above and below this range. Therefore the effects of plant nutrients on scab often are influenced by the soil pH and, in some cases, the soil pH is affected by the amendments. In addition, soil pH has a significant effect on the availability of macro- and micronutrients. Unfortunately, the soil pH at the tuber surface usually has not been assessed.

SULFUR

As early as 1897 Wheeler and Adams (64) demonstrated that applications of elemental sulfur (S) to soil would control potato scab. This is one of the first examples of control of a soilborne pathogen with a soil nutrient. Though other mechanisms have been proposed for the suppressive effect of S on scab, most studies indicate that disease suppression is due to a reduction of soil pH (3,4,28,39,40,41,42,50,58) which occurs when S is oxidized. Disease is usually much less severe in soils below pH 5.4 than above this pH value. Disease suppression provided by S is reversed by the application of lime (34,48).

Growth of *S. scabies* appears to be directly affected by pH. Waksman (60) showed that growth of *S. scabies* was inhibited in nutrient media adjusted to pH 4.9 - 5.2. Optimum growth in sterile soil was between pH 5.0 and 8.0, but strains varied in their response to pH. Loria *et al.* (35) also showed that growth of *S. scabies* was inhibited below pH 4.8 but that another *Streptomyces* sp., also pathogenic on potato, could grow as low as pH 4.0, the lowest pH tested. Elemental S is not toxic to *S. scabies* when incorporated into agar (28). However, hydrogen sulfide, which is produced from S in soil under anaerobic conditions, inhibits the pathogen in pure culture (59).

Davis *et al.* (14) demonstrated that both elemental S and gypsum $(CaSO_4 \cdot H_2O)$ reduced scab and soil pH when applied to soil which had a pH of 7.7. However, they indicated that the reduction in soil pH was not sufficient to account for disease suppression. Calcium (Ca) levels in tuber peelings from S-treated plots were lower than from control plots, suggesting that the reduction of disease in S-treated plots was indirect and due to a decrease in Ca in tuber peelings. As mentioned subsequently (15,16), the concentration of Ca in tuber periderm tissue is thought to be positively correlated with susceptibility to scab. Unfortunately, Ca levels in the tuber periderm were not monitored in most studies in which disease suppression by S was demonstrated.

Application of S for control of potato scab is practiced commercially, albeit to a limited extent. The use of S for scab control is often impractical due to the difficulty of lowering the pH of some soils. In addition, potatoes are often rotated with crops which are intolerant of low soil pH.

CALCIUM AND POTASSIUM

When Ca is added to soil in the form of calcium carbonate ($CaCO_3$, lime) the incidence and severity of potato scab (8,17,22,50,58,64) is increased in proportion to the increase in soil pH. Blodgett and Cowan (8) showed that addition of CaO, $CaCO_3$, $CaSO_4$ and Na_2CO_3 had no affect on disease independent of soil pH effects. When $Ca(OH)_2$ was applied to soil to adjust the soil pH to 5.0-9.0, there was a positive correlation between disease occurrence and soil pH. Research conducted in solution-culture indicated that there was no relationship between Ca content of the tubers and scab susceptibility (30).

However, there is evidence for a direct effect of Ca on the susceptibility of potato tubers to *S. scabies*. Horsfall *et al.* (29) suggested that an increase in available Ca in the soil results in higher Ca levels in the tuber, which increases susceptibility to *S. scabies*. Davis and co-workers (15,16) showed that severity of potato scab was correlated with Ca concentration in the peelings of tubers grown in soil with high exchangeable Ca. They also noted that low soil moisture, which increases scab, also resulted in higher Ca concentrations in tuber peelings. Goto (22) found that scab ratings were better correlated with exchangeable Ca, at levels above 150 mg Ca/100 g soil, than with soil pH in volcanic ash soil. Further, Mortvedt *et al.* (46) showed that copper sulfate ($CuSO_4$) applications, which are known to inhibit Ca uptake, provide some control of potato scab. However, as mentioned subsequently, copper (Cu) is toxic to *S. scabies* (6).

Studies on the ratio of Ca and potassium (K) concentrations in the tuber periderm have produced conflicting findings. Davis *et al.* (15) found that scab severity increased as the Ca:K ratio in potato tubers increased. However, Doyle and MacLean (18) showed that when Ca:K ratios were manipulated independently of soil pH, they did not affect scab ratings.

A few studies have evaluated the effect of K independent of Ca. Those results have consistently shown that K applications and K concentrations in the tubers do not influence scab incidence or severity (25,52,62).

NITROGEN

Nitrogen (N) appears to influence potato scab indirectly through its influence on soil pH. Ammoniacal forms of N, e.g. ammonium sulfate [$(NH_4)_2SO_4$], and those which produce ammonia (NH_3) when added to soil, e.g. urea, can acidify the soil through two processes. The primary mechanism is through nitrification: H ions are released to the soil during the conversion of ammonium (NH_4) to nitrite. Secondarily, as (NH_4) ions are absorbed by roots, H ions are exuded to maintain electroneutrality. Ammonia is toxic to S. scabies in vitro (2), but formation of NH_3 from NH_4 fertilizers is unlikely to occur in soil (11). Nevertheless, the possibility of a direct effect of NH_4 ions or NH_3 on scab cannot be ruled out.

Applications of $(NH_4)_2SO_4$ have reduced both the soil pH and the scab index in most (20,49,58), but not all (17,23) studies. Dippenaar (17), for instance, reduced soil pH to 5.0-5.5 with $(NH_4)_2SO_4$ but observed only a slight reduction in scab. Ammonium sulfate did not control scab in Scotland (23), but soil pH was not reported. In early studies, Wheeler (64,65) demonstrated that scab was less severe with $(NH_4)_2SO_4$ than with Na-NO_3 fertilization. Reichard & Wenzl (52) confirmed the inhibitory effect on scab of NH_4 relative to nitrate (NO_3) (using neutralized (NH_4NO_2) and also showed that a significantly lower soil pH resulted from applications of $(NH_4)_2SO_4$. Neutralized (NH_4NO_2) did not control potato scab (9). Davis et al. (15) found that the scab index was not significantly affected by ($Ca(NO_3)_2$ or $(NH_4)_2SO_4$ applied to soil at pH 7.5-8.0.

The level of N does not appear to directly affect potato scab: there was no correlation between scab and the NO_3 content of tuber periderm (15) or total soil N in commercial fields (53). Further, N fertilization was positively associated with scab in a multiple regression model for only one of three years (25). Scab was increased (33,62) or not affected (52) by increasing amounts of neutralized (NH_4^+). High N applications which delay tuberization can indirectly affect scab. Lapwood and Dyson (33) deduced that tubers on plants receiving high levels of N matured later, when the soil moisture was lower and therefore more favorable for infection by S. scabies, than the no N controls.

PHOSPHORUS

Phosphorus (P) seems to have little direct effect on potato scab. However, the form of P may influence disease if the soil pH is altered, as occurs with applications of basic slag [(CaO)$_5$.P$_2$O$_5$.SiO$_2$], a phosphate fertilizer used in Europe. After repeated slag applications for 4 to 15 years, both soil pH and the scab index were increased in four of five experiments (52,62,63). In a comparison of superphosphate, hyperphosphate, and basic slag at the rate of 100 kg P$_2$O$_5$/ha over a 14-year period (63), basic slag significantly increased the soil pH and the scab index compared to hyperphosphate, superphosphate, and the no P control.

These phosphate fertilizers have varying effects on soil pH as they differ in the percentage of calcium oxide (CaO) equivalents and the form of Ca (63). Basic slag has 45% CaO as CaO and Ca$_2$SiO$_4$, both of which are fairly readily available forms of Ca. Hyperphosphate has 48% CaO, mostly as carbonate apatite, a relatively insoluble material. Although superphosphate is soluble, it is quickly converted to less available forms of P, and thus has little effect on soil pH.

In contrast to results with other forms of P, Davis et al. (15) obtained a significant reduction in scab incidence with triple-superphosphate in two years, although differences in disease incidence between the levels of phosphate applied were not statistically significant. Phosphate content of tuber peelings was negatively correlated with the scab severity index. The reduction in disease did not appear to be due to toxicity of triple-superphosphate to S. scabies, since triple-superphosphate did not inhibit S. scabies in vitro. The authors concluded that triple-superphosphate indirectly affected scab by reducing the Ca:K ratio of the tubers.

COPPER

Copper (Cu), in the form of CuSO$_4$, has been used for scab control. Prior to 1900, CuSO$_4$ was among the most effective compounds tested for scab control when used as a tuber dip (7) or applied to the soil (27). Mortvedt et al. (46) confirmed that CuSO$_4$ significantly reduced scab when 56 kg/ha was applied to soil with a pH of 5.5, but this rate was ineffective in soil at pH 4.9. However, scab indices were lower in the more acidic soil and the effect of pH may have masked the effect of Cu.

In other studies (31,45,49) $CuSO_4$ was ineffective for scab control, but the soil pH was not reported. Cuprous oxide reduced scab in soils of pH 5.2-5.8 (31), although the reduction was not statistically significant. The EDTA-soluble Cu in soil was negatively correlated with scab in a multiple regression model for one of three years (25).

Copper appears to be toxic to *S. scabies*. Copper, at a concentration (100 ppm) comparable to that found in some unamended soils (11), reduced the number of *S. scabies* colonies recovered from sand-perlite medium as well as the scab index in a greenhouse experiment (6). However, Cu levels in potato plants do not appear to be related to disease. In another experiment in which only the roots of potato plants were exposed to Cu, there was no reduction in scab symptoms on tubers (46). In addition, under field conditions, the Cu content of tuber periderm was not correlated with the scab index (15).

Although Cu does control potato scab, it is not used commercially because of its negative effects on plant growth and yield. Some of the Cu concentrations tested were phytotoxic to potato, resulting in poor root growth (6,46), delayed tuber bulking (36), and yield reductions (27,31).

IRON

A few studies have been done to examine relationships between iron (Fe) and scab control. Ferrous sulfate applied to soil did not control scab in one field trial (49), but as a tuber dip with scab-free seed it reduced scab incidence from 47% to 18% (7). The concentration of EDTA-soluble Fe in soil (25) and the Fe content of tuber peelings (15) have been negatively correlated with scab. In addition, Fe deficiency inhibits suberization in bean roots (57), and suberized lenticels on potato tubers are resistant to *S. scabies* (1). However, no data are available on the relationship between Fe concentration and suberization of potato tuber lenticels.

ZINC

Several studies have demonstrated that zinc (Zn) does not control potato scab. Zinc (31), ZnO (31), and $ZnSO_4$ (49) were ineffective in field trials. The Zn content of tuber peelings ($ZnSO_4$) was not correlated with the scab index (15). This lack of disease control may be due to the tolerance of *S. scabies* to Zn. The number of *S. scabies* colonies was reduced only 18% by 1000 ppm Zn (6). However, application of $ZnSO_4$ to soil at pH 6.8-7.0 resulted in a slight but statistically significant increase in the percentage of marketable tubers in one study (24).

BORON

Studies on the application of boron (B) for scab control are inconclusive. Boron significantly, albeit slightly, reduced scab in a one-year trial (52), whereas in one of three trials, borax ($Na_2B_4O_7.10H_2O$) significantly increased scab (19). Soil applications of borax had no effect in two other studies (24,31). There was no correlation between the B content of tuber peelings and the scab index (15).

MANGANESE

Manganese (Mn) has been shown to control scab, but there are unexplained inconsistencies in the efficacy of this micronutrient. A comparison of two studies illustrates this point. Each study was conducted at four locations in Scotland, where the soil pH ranged from 6.2 to 7.0 before treatment. McGregor and Wilson (44) found that 31 kg manganese sulfate ($MnSO_4$)/ha banded in the furrow significantly reduced the scab index. Control was achieved over a range of pretreatment soil Mn contents of 87 to 1,007 ppm. Gilmour *et al.* (21) obtained no significant reduction in scab incidence with 57 kg $MnSO_4$/ha spread in the furrow. Other field studies have confirmed both the reduction in potato scab due to Mn applications (24,43,46) and the lack of disease control (3,47,49,52,54). When changes in soil pH due to Mn applications were monitored, there was no effect on soil pH (3,46,51) except in one study (21). Gilmour *et al.* (21) observed a reduction in pH of 0.3 to 0.8 pH unit; however the lowest pH was 5.9, which was not low enough to reduce the scab severity.

Both the rate of Mn application and its placement in the planting furrow seem to influence efficacy. For example, ≤168 kg $MnSO_4$/ha banded in the furrow was ineffective; however both 168 kg/ha spread in the furrow and 504 kg/ha banded in the furrow controlled scab (46). Other studies demonstrated that spreading ≤125 kg $MnSO_4$/ha in the furrow was ineffective (3,21,54). The results of McGregor and Wilson (43,44) are an exception, since they obtained control with low amounts of banded $MnSO_4$.

The influence of Mn distribution on disease control suggests that Mn can be toxic to *S. scabies*. Studies conducted by Mortvedt *et al.* also indicate that direct toxicity to *S. scabies* may be involved in disease suppression. When inoculum was mixed with Mn solutions before application to potatoes in a greenhouse experiment, ≥ 2 ppm Mn significantly reduced scab (47). In another experiment in which only the roots of potatoes were exposed to Mn, there was no reduction in scab (46). Further, Mn (250 ppm) incorporated into agar inhibited growth but not sporulation of *S. scabies* (46). Under anaerobic conditions Mn is found in the soil solution at concentrations comparable to those which inhibit the pathogen (10).

A relationship between the concentration of Mn in tuber peelings and scab severity has been suggested. The Mn content of peelings was negatively correlated with the scab index within the ranges of 36-70 ppm (15) and 26-120 ppm (46). Likewise, the Mn content of the peelings was significantly increased by Mn treatments which also significantly reduced scab (43,44,46,47). However, the scab severity associated with low levels of Mn was not consistent in these studies. For example, the scab index was 46 (on a scale of 0 to 100) with a Mn content of 120 ppm (46), and 8.2 with 110 ppm (47). Barnes and McAllister (5) observed a significant increase in the Mn content of tuber peelings without a corresponding decrease in scab.

The results of field trials apparently have not been favorable enough to warrant the use of $MnSO_4$ for scab control. Further, Mn can be toxic to potato. A significant yield reduction was observed when 125.5 kg $MnSO_4$/ha was applied directly to the seed tubers (3). In addition, potatoes are often grown on sandy soils, which are relatively well-aerated. Hence, reducing conditions necessary to solubilize Mn may not prevail in enough microsites or for a long enough period to make Mn effective.

SUMMARY AND CONCLUSIONS

We have reviewed research on the effects of ten plant nutrients on potato scab. In many cases, inadequate information on edaphic factors, uptake of nutrients by host plants, and population dynamics of *S. scabies* limited the value of the information generated. Lack of appropriate statistical analyses also was a problem in evaluating the results. Therefore, the following suggestions are made for future research on control of potato scab with plant nutrients.

Accurate information on soil pH is essential for any study of potato scab, since direct effects of pH on *S. scabies* and nutrient availability have been clearly demonstrated. Soil pH is spatially variable; sufficient measurements should be made to accurately account for this variation. Determining the pH at the tuber surface, where infection occurs, would help to better understand the influence of pH on the infection process.

The concentration of nutrients in the potato periderm also should be measured, in order to evaluate the effect of nutrient additions on the physiology of the host. Davis *et. al.* (15,16) hypothesized that the effects of various control measures which reduced scab could be explained by the resulting increase in the Ca level of the potato plant. This proposed stimulation of susceptibility by Ca should be verified. In addition, Cu, B, Fe (38), and Mn (26) are involved in lignin biosynthesis in other crop plants. Their role in suberization and lignification in the potato tuber and resultant effects on scab susceptibility should be examined.

The fluctuations with time of soil pH and nutrient concentrations in soil also are important, since the susceptibility of the host is limited in time. Infection only occurs as the stomates in the immature tubers are transformed into lenticels. Once the lenticels have matured and are completely suberized, the pathogen is unable to gain ingress (1). The pH or nutrient concentrations measured at harvest may not reflect the conditions present during the infection period, and consequently may have little to do with the disease incidence or severity. Therefore measurements should be taken for the period (approximately six weeks) during which tuber initiation and expansion occur.

Research on *S. scabies* and potato scab is limited by inadequate techniques for accurately measuring the population density of this pathogen in soil. Information on the population dynamics of this actinomycete would provide an additional criterion by which to evaluate control strategies. Although a variety of media and techniques are available for isolating actinomycetes from soil (32), there is no selective agent for separating *S. scabies* from the many other *Streptomyces* spp. recovered. Furthermore, there is no rapid and reliable method to test the pathogenicity on potato of isolates tentatively identified as *S. scabies*.

A related problem is the need for verification of the species used in studies on potato scab. Several species of *Streptomyces* have been reported to cause various types of scab symptoms (13,37,56). Species identification cannot be based on symptoms, and different pathogenic species respond differently to edaphic factors (56). Unfortunately, according to Bergey's Manual of Determinative Bacteriology (51), the taxonomic position of *S. scabies* is uncertain. A world-wide collection of pathogenic isolates must be examined in detail in order to validly describe this species, and to find characteristics useful in identification.

The effects of micronutrients on the growth and survival of *S. scabies* may be important in managing scab. Toxic levels of Cu (6) and Mn (47) have provided control of scab in greenhouse experiments. Perhaps micronutrient toxicity is one of the mechanisms providing control of scab at low soil pH. These observations, as well as the tolerance of *S. scabies* to Zn (6), should be verified under field conditions. McGregor and Wilson (44) hypothesized that Mn toxicity was the mechanism for control of scab provided by irrigation, S applications, and green manures, measures which could increase the level of available Mn in the soil. This hypothesis should be examined further.

Greenhouse and field experiments should be designed so that correct statistical analyses of the data are possible. Disease incidence should be weighted to account for the number of tubers actually examined. Disease severity values expressed as percentage of the tuber surface scabbed often should be transformed by the arcsin-square root transformation when the percentages are low or cover a wide range of values. Those rating scales which are not suited to parametric analyses, e.g. scab indices, should be treated with non-parametric techniques, such as categorical data modeling (55). The interaction of pH and nutrient availability also should be taken into consideration in the analysis of the data. One way to do this is to use pH as a covariate in the analysis of variance, so that all comparisons between treatments (nutrient additions, for example) are weighted based on pH. Based on the literature, the most promising control strategy is adjusting the soil pH to a level at which the activity of the pathogen is reduced or inhibited. Using ammoniacal forms of N may be the most effective way of obtaining this pH reduction without creating an unfavorable pH for rotation crops. This is an example of using the interaction between soil pH and plant nutrients as an aid in control. Application of $MnSO_4$ would be another way to exploit this interaction beneficially to achieve a slight reduction in soil pH, while possibly providing additional control through inhibition of the pathogen by Mn toxicity. The effectiveness of $MnSO_4$ should be tested further in field trials. Combinations of nutrients, especially micronutrients, have not been extensively examined; there may be combinations of nutrients which are more effective than single applications.

Since *S. scabies* is a ubiquitous, well-adapted soil inhabitant, it is very unlikely that a single control strategy will provide total disease control. Moreover, elimination of scab may be both unfeasible and unnecessary, since growers, processors, and consumers are willing to tolerate a low level of scab. In addition to the use of plant nutrients to manage common scab, other control methods include seed treatments, crop rotation and green manure crops, irrigation timing, and varietal resistance. The latter two methods are most effective, but only varietal resistance gives reliable control. An integrated disease management program, including plant nutrients, may be the most effective way to achieve adequate disease reduction.

LITERATURE CITED

1. Adams, M.J. 1975. Potato tuber lenticels: development and structure. Ann. Appl. Biol. 79:265-273.
2. Afanasiev, M.M. 1937. Comparative physiology of actinomycetes in relation to potato scab. Neb. Agric. Exp. Sta. Res. Bull. 92. 63 pp.
3. Barnes, E.D. 1972. The effects of irrigation, manganese sulphate and sulphur applications on common scab of the potato. Rec. Agric. Res. Minist. Agric. Nth. Ire. 20:35-44.
4. Barnes, E.D., and Chestnutt, D.M.B. 1966. Some aspects of the control of common scab in potatoes. Rec. Agric. Res. Minist. Agric. Nth. Ire. 15:35-42.
5. Barnes, E.D., and McAllister, J.S.V. 1972. Common scab of the potato: the effects of irrigation, manganese sulphate, and sulphur treatments for common scab of the potato on the mineral composition of plant material and soil extracts. Rec. Agric. Res. Minist. Agric. Nth. Ire. 20:53-58.
6. Basu Chaudhary, K.C. 1967. Influence of copper and zinc on the incidence of potato scab. Neth. J. Plant Path. 73:49-51.
7. Beckwith, M.H. 1887. Report of the assistant horticulturist. N.Y. (State) Agric. Exp. Stn. Ann. Rep. 6:307-315.
8. Blodgett, F.M., and Cowan, E. K. 1935. Relative effects of calcium and acidity of the soil on the occurrence of potato scab. Am. Potato J. 12:265-274.
9. Böhmig, H.J., Friessleben, G., Gerdes, K., Truckenbrodt, M., Janke, C., Lücke, W., and Schnieder, E. 1975. Einfluß hoher mineralischer Stickstoffdüngung und Beregnung auf Ertrag und Qualität der Kartoffel. 2. Mitteilung: Einige Qualitätsmerkmale der Knollen, Auftreten bedeutsamer Pilz-und Bakterienkrankheiten, Lagerfähigkeit des Ernteguts sowie Erosion der Kartoffeldämme. Arch. Acker-u. Pflanzenbau u. Bodenkd. 19:793-809.
10. Bohn, H.L., McNeal, B.L., and O'Connor, G.A. 1979. Soil Chemistry. John Wiley & Sons, New York. 329 pp.
11. Brady, N.C. 1974. The Nature and Properties of Soils, 8th ed. Macmillan Publishing Co., Inc., New York. 639 pp.

12. Cetas, R.C., and Sawyer, R.L. 1962. Evaluation of uracide for the control of common scab of potatoes on Long Island. Am. Potato J. 39:456-459.

13. Corbaz, R. 1964. Étude des streptomycètes provoquant la gale commune de la pomme de terre. Phytopathol. Z. 51:351-361.

14. Davis, J.R., Garner, J.G., and Callihan, R.M. 1974. Effects of gypsum, sulfur, terraclor and terraclor Super-X for potato scab control. Am. Potato J. 51:35-43.

15. Davis, J.R., McDole, R.E., and Callihan, R.H. 1976. Fertilizer effects on common scab of potato and the relation of calcium and phosphate-phosphorus. Phytopathology 66:1236-1241.

16. Davis, J.R., McMaster, G.M., Callihan, R.H., Nissley, F.H., and Pavek, J.J. 1976. Influence of soil moisture and fungicide treatments on common scab and mineral content of potatoes. Phytopathology 66:228-233.

17. Dippenaar, B.J. 1933. Environmental and control studies of the common scab disease of potatoes caused by *Actinomyces scabies* (Thaxt.) Güss. Union S. Africa Dept. Agric. Sci. Bull. 136. 78 pp.

18. Doyle, J.J., and MacLean, A.A. 1960. Relationships between Ca: K ratio, pH, and prevalence of potato scab. Can. J. Plant Sci. 40:616-619.

19. van Emden, J.H., and Labruyère, R.E. 1958. Results of some experiments on the control of common scab of potatoes by chemical treatment of the soil. Eur. Potato J. 1:14-24.

20. Emilsson, B., and Gustafsson, N. 1954. Studies on the control of common scab on the potato. Acta Agric. Scand. 4:33-62.

21. Gilmour, J., Crooks, P., Rodger, J.B.A., Wynd, A., and Mackay, A.J.M. 1968. Manganese soil treatment for the control of common scab of potatoes. Edinburgh Sch. Agric. Exp. Work 1967:36- 37.

22. Goto, K. 1985. Relationships between soil pH, available calcium and prevalence of potato scab. Soil Sci. Plant Nutr. 31:411-418.

23. Gray, E.G., Smith, J.D., and Knight, B.C. 1961. Some effects of soil treatments for control of common scab and black scurf of potato. Eur. Potato J. 4:277-278.

24. Guntz, M., and Coppenet, M. 1957. Essais de traitements contre la gale commune de la pomme de terre. Phytiat. -Phytopharm. 6:187-195.

25. Gusenleitner, J. 1974. Der Zusammenhang zwischen ökologischen bzw. betriebswrtschaftlichen Gegebenheiten und dem Befall mit Kartoffelschorf (*Streptomyces scabies* und *Spongospora subterranea*). Bodenkultur 25:63-74.

26. Halliwell, B. 1978. Lignin synthesis: the generation of hydrogen peroxide and superoxide by horseradish peroxidase and its

stimulation by manganese (II) and phenols. Planta 140:81-88.

27. Halsted, B.D. 1895. Field experiments with potatoes. N. J. Agric. Exp. Stn. Bull. 112. 20 pp.

28. Hooker, W.J., and Kent, G.C. 1950. Sulfur and certain soil amendments for potato scab control in the peat soils of northern Iowa. Am. Potato J. 27:343-365.

29. Horsfall, J.G., Hollis, J.P., and Jacobson, H.G.M. 1954. Calcium and potato scab. Phytopathology 44:19-24.

30. Houghland, G.V.C., and Cash, L.C. 1956. Some physiological aspects of the potato scab problem. II. Calcium and calcium-Potato J. 33:235-241.

31. Ken Knight, G. 1941. Studies on soil actinomycetes in relation to potato scab and its control. Mich. Agric. Exp. Stn. Tech. Bull. 178. 48 pp.

32. Kutzner, H.J. 1981. The family Streptomycetaceae. Pages 2028-2090 in: The Prokaryotes: A Handbook on Habitats, Isolation and Identification of Bacteria, Vol. 2, M.P. Starr, H. Stolp, H.G. Trüper, A. Balows, and H.G. Schlegel, eds. Springer-Verlag, Berlin.

33. Lapwood, D.H., and Dyson, D.W. 1966. An effect of nitrogen on the formation of potato tubers and the incidence of common scab (*Streptomyces scabies*). Plant Pathology 15:9-14.

34. Larson, R.H., Albert, A.R., and Walker, J.C. 1938. Soil reaction in relation to potato scab. Am. Potato J. 15:325-330.

35. Loria, R., Kempter, B.A., and Jamieson, A.A. 1986. Characterization of *Streptomyces* spp. causing scab in the Northeast. (Abstr.) Am. Potato J. 63:440.

36. Mader, E.D., and Blodgett, F.M. 1935. Potato spraying and potato scab. Am. Potato J. 12:137-142.

37. Manzer, F.E., McIntyre, G.A., and Merriam, D.C. 1977. A new potato scab problem in Maine. Univ. Maine Tech. Bull. 85. 24 pp.

38. Marschner, H. 1986. Mineral Nutrition of Higher Plants. Academic Press, London. 674 pp.

39. Martin, W.H. 1920. The relation of sulphur to soil acidity and to the control of potato scab. Soil Sci. 9:393-408.

40. Martin, W.H. 1931. Soil reaction and the potato crop. Am. Potato J. 8:59-64.

41. McAlister, J.S.V. 1971. The use of sulphur to control common scab of potatoes. Res. Exp. Rec. Minist. Agric. Nth. Ire. 11:111-114.

42. McCreary, C.W.R. 1967. The effect of sulphur application to the soil in the control of some tuber diseases. Proc. 4th Brit. Insect. Fungic. Conf. Brighton 1:303-308.

43. McGregor, A.J., and Wilson, G.C.S. 1964. The effect of

applications of manganese sulphate to a neutral soil upon the yield of tubers and the incidence of common scab in potatoes. Plant Soil 20:59-64.

44. McGregor, A.J., and Wilson, G.C.S. 1966. The influence of manganese on the development of potato scab. Plant Soil 25:3-16.

45. Meyer, C. 1940. Eenige resultaten van proeven en waarnemingen over het optreden van aardappelschurft. Tijdschr. Plantenziekt. 46:19-26.

46. Mortvedt, J.J., Berger, K.C., and Darling, H.M. 1963. Effect of manganese and copper on the growth of *Streptomyces scabies* and the incidence of potato scab. Am. Potato J. 40:96-102.

47. Mortvedt, J.J., Fleischfresser, M.H., Berger, K.C., and Darling, H.M. 1961. The relation of soluble manganese to the incidence of common scab in potatoes. Am. Potato J. 38:95-100.

48. Muncie, J.H., Moore, H.C., Tyson, J., and Wheeler, E.J. 1944. The effect of sulphur and acid fertilizer on incidence of potato scab. Am. Potato J. 21:293-304.

49. Mygind, H. 1962. Forsøg med bekaempelse af kartoffelskurv og rodfiltsvamp. Tidsskr. Plavl. 66:423-457.

50. Odland, T.E., and Albritton, H.G. 1950. Soil reaction and calcium supply as factors influencing the yield of potatoes and the occurrence of scab. Agron. J. 42:269-275.

51. Pridham, T.G., and Tresner, H.D. 1974. Family VII. Streptomycetaceae Waksman and Henrici 1943, 339. Pages. 747-829 in: Bergey's Manual of Determinative Bacteriology, 8th ed., R. E. Buchanan and N.E. Gibbons, eds. The Williams & Wilkins Company, Baltimore.

52. Reichard, T., and Wenzl, H. 1976. Beiträge zu Düngung und Kartoffelschorf. Pflanzenschutzberichte 45:57-69.

53. Richardson, J.K., and Heeg, T.J. 1954. Potato common scab investigations. I. A survey of disease incidence in southern Ontario. Can. J. Agric. Sci. 34:53-58.

54. Rodger, J.B.A., Wynd, A., and Gilmour, J. 1967. Manganese and sulfur treatments for the control of common scab of potatoes. Edinburgh Sch. Agric. Exp. Work 1966:24-25.

55. SAS Institute Inc. 1985. SAS User's Guide: Statistics, Version 5 Edition. SAS Institute Inc., Cary, NC. 956 pp.

56. Scholte, K., and Labruyère, R.E. 1985. Netted scab: a new name for an old disease in Europe. Potato Res. 28:443-448.

57. Simons, P.C., Kolattukudy, P.E., and Bienfait, H.F. 1985. Iron deficiency decreases suberization in bean roots through a decrease in suberin-specific peroxidase activity. Plant Physiol. 78:115-12.

58. Terman, H.L., Steinmetz, F.H., and Hawkins, A. 1948. Effects of

certain soil conditions and treatments upon potato yields and the development and control of potato scab. Maine Agric. Exp. Stn. Bull. 463. 31 pp.

59. Vlitos, A.J., and Hooker, W.J. 1951. The influence of sulfur on populations of *Streptomyces scabies* and other streptomycetes in peat soil. Am. J. Bot. 38:678-683.

60. Waksman, S.A. 1922. The influence of soil reaction upon the growth of actinomycetes causing potato scab. Soil Sci. 14:61-79.

61. Wenzl, H. 1975. Die Bekämpfung des Kartoffelschorfes durch Kulturmaßnahmen. Z. Pflanzenkr. Pflanzenschutz 82:410-440.

62. Wenzl, H., and Reichard, T. 1974. Der Einfluß von Mineraldüngern auf Kartoffelschorf (*Streptomyces scabies* [Thaxt.] Waksman et Henrici und *Spongospora subterranea* [Wallr.] Lagerh.). Bodenkultur 25:130-137.

63. Wenzl, H., Reichard, T., and Gusenleitner, J. 1972. Kalkzustand des Bodens und Befall der Kartoffel mit Actinomyces-Schorf (*Streptomyces scabies* [Thaxt.] Waksman et Henrici) und Pulverschorf (*Spongospora subterranea* [Wallr.]Lagerh.). Bodenkultur 23:227-251.

64. Wheeler, H.J., and Adams, G.E. 1897. On the use of flowers of sulfur and sulfate of ammonia as preventive of the potato scab in contaminated soils. Rhode Is. Expt. Stn. Ann. Rept. 10:254-268.

65. Wheeler, H.J., and Tucker, G.M. 1895. Upon the effect of farmyard manure and various compounds of sodium, calcium and nitrogen upon the development of potato scab. Rhode Is. Agric. Exp. Stn. B 33:51-79.

ROLE OF NUTRITION IN
DISEASES OF COTTON

Alois A. Bell
USDA-ARS, Southern Crops Research Laboratory
College Station, TX 77841

The annual production of cotton is about 13 million bales (500 lb each) in the USA and 60-70 million bales worldwide (107). Cotton is grown in all of the southern states in the USA, but over 80% of the total USA yield comes from Texas, California, Mississippi, Louisiana, Arkansas, and Arizona. The average yield is 3 bales/ha in the USA. The availability of H_2O and N, along with damage by pests, are the major limits on cotton yields. About 10% of the potential yield in the USA is lost to diseases and nematodes and 17% to insects. The losses from individual diseases and nematodes since 1956 are summarized in Table 1. These data are adapted from the cotton disease loss estimates of the Cotton Disease Council, published annually in the Proceedings of the Beltwide Cotton Production Research Conferences, National Cotton Council, Memphis, Tennessee.

All major cotton diseases, except bacterial blight and boll rots, are caused by soilborne pathogens. Seedling diseases are caused primarily by the fungi, *Pythium ultimum* Trow., *Thanatephorus cucumeris* (Frank) Donk (imperfect stage *Rhizoctonia solani* Kuehn), and *Chalara elegans* Nag Raj and Kendrick [syn. *Thielaviposis basicola* (Berk and Br.) Ferr.]. The root-knot nematode, *meloidogyne incognita* Kofoid & White, and the reniform nematode, *Rotylenchulus reniformis* Linford & Oliviera, generally cause the greatest losses from nematodes. All of the soilborne pathogens invade the root or that portion of stem below the soil line. These pathogens have only one or a few reproductive cycles per year, and then survive as dormant

This chapter is in the public domain and not copyrightable. It may be freely reprinted with customary crediting of the source. The American Phytopathological Society, 1989.

167

propagules in the soil until a host plant is present. Consequently, soil conditions strongly affect both invasion of the host and survival of the pathogen.

Table 1. Losses from potential production of cotton caused by diseases and nematodes in the USA from 1956 through 1988.[a]

Diseases	Losses from potential production			
	1956-1965	1966-1975	1976-1985	1986-1988
	------------------ (%) ------------------			
Seedling diseases	2.74	3.28	2.67	2.10
Verticillium wilt	2.17	2.95	2.16	1.39
Fusarium wilt	1.05	0.77	0.35	0.34
Phymatotrichum root rot	1.32	0.65	1.38	0.34
Bacterial blight	1.50	1.04	0.23	0.07
Boll rots	2.14	3.06	2.07	2.91
Nematodes	1.47	2.19	1.85	1.95
Total	12.39	13.94	10.71	9.10

[a] Data adapted from the cotton disease loss estimates of the Cotton Disease Council, published annually in the Proceedings of the Beltwide Cotton Production Research Conferences.

This chapter principally deals with how macro- and microelements affect cotton diseases. However, it must be remembered that each element has strong interactions with other elements, soil moisture, temperature, agricultural chemicals, and microorganisms. In some cases these interactions will be discussed in order to better understand how elements affect disease.

EFFECTS OF NUTRIENTS ON BIOTIC DISEASES

Seedling Diseases

Nutrient elements can affect disease severity by changing the inoculum potential or virulence of the pathogen, or by changing the resistance of the host to the pathogen. Preplant fertilization with anhydrous NH_3 and (NH_4^+) salts reduces populations of pathogens such as *Phythium ultimum* (154), *Fusarium* spp. (149), and *Sclerotium rolfsii* Sacc. (7). The suppressive effects are due to both toxic effects of NH_3 and inhibition of saprophytic growth. Ammonia is toxic to *P. ultimum, Fusarium* spp., and *Rhizoctonia solani* (50,51,52,149,154). Combining solarization (covering moist soil with clear plastic sheets for several weeks to allow heating by sunlight) with $(NH_4)_2HPO_4$ or $(NH_4)_2SO_4$ amendments can reduce *P. ultimum* to nondetectable levels (154). Solarization might increase the production, diffusion, and retention of NH_3 in soils.

Manganese is toxic to *R. solani* in culture media at 60 mg/L (141). In addition, the incidence of disease from this pathogen is reduced in Mn-toxic soils compared to the same soils treated with lime to remove the toxicity.

Elemental amendments may change populations of beneficial organisms in soil, and the beneficial organisms, in turn, may change the availability or form of the amendments. Fertilization with K increases and with N decreases populations of the *Penicillium funiculosum* series, which are antagonistic to pathogens such as *Fusarium* spp., *Verticillium albo-atrum* Reinke & Berth., and *R. solani*. The *Penicillium oxalicum* series, other antagonists of these pathogens, develop greater populations in neutral soils (pH 6.5 & 7.3) than in acid soils (pH 4.9 & 5.4) (65). The biocontrol organisms *Pseudomonas fluorescens* (Trevisan) Migula and *Enterobacter cloacae* (Jordan) Hormache and Edwards suppress *P. ultimum and R. solani* in culture by producing NH_3 by *E. cloacae* in the soil amendment of soils with organic N, e.g., manure or asparagine, might be used to enhance effectiveness of NH_3-producing biocontrol agents, providing that NH_3 toxicity to the cotton plant can be avoided. Total numbers of bacteria and fungi (including antagonists) in the rhizosphere vary significantly with fertilizer treatments and cultivars (16).

The pathogenic capabilities of *R. solani* to cotton depend on the nutritional status of the inoculum, and adding N amendments to soil may increase the pathogenicity of this fungus. The mean lesion area on cotton hypocotyls at 48-60 hours after inoculation was 78 and 7 mm^2 when the mycelium used as inoculum was grown on media containing 2.0 and 0.5 g/L of asparagine, respectively (168). Mycelium growing from inoculum deficient in N rapidly absorbs N-containing compounds from the external environment and utilizes these compounds for pathogenic activities (169). Thus, adding $(NH_4)_2PO_4$ or KNO_3 to soil results in an increase in lesion development.

In their review of germination and stand-establishment in cotton,

Christiansen and Rowland (27) concluded that Ca is the most critically required exogenous nutrient element for the initial development of the cotton plant. In addition to naturally low concentrations of Ca in soils, various sources of stress, such as chilling (70,117), Al toxicity (57), salinity (66), and drouth (116), also may induce or aggravate Ca deficiency. Adding P and N fertilizers to soil further enhances Ca deficiency (2,117). Infection with root-knot nematodes for 6 days lowered Ca concentrations in leaves, but not in stems or roots (22). When Ca is deficient, membrane integrity is lost, and root cells in the region of elongation collapse and become necrotic. Then roots are unable to elongate and penetrate into surrounding soil (48,116). Collapse and death of portions of the hypocotyl and the terminal bud, along with chlorosis and necrosis of the cotyledons, may occur with severe Ca deficiency (172). The translocation of carbohydrates from the leaves to the stem and roots also is disrupted (58). Consequently, carbohydrates accumulate in the leaf, and concentrations lower than normal occur in the stem and root, probably contributing to the stunting that is associated with Ca deficiency. The symptoms of Ca deficiency closely resemble symptoms that normally are attributed entirely to diseases incited by soilborne fungi (172).

Amendments with Ca generally increase the percentages of seedling emergence, improve seedling vigor, and decrease the incidence of disease, if soils are low in Ca, of if stresses that aggravate Ca deficiency are present. Beneficial effects of Ca amendments have been demonstrated for the nutrient-depleted soils of Georgia (179), the acidic Mn-toxic soils of Alabama (129), the acidic soils of East Texas (43,122), and saline sand-nutrient systems (66). Batson (11) found that Ca, Mg, and K concentrations in seed exudates of 31 strains and five cultivars of *G. hirsutum* correlated negatively with pre-emergence damping-off, and positively with final stand and yield. Thus, breeding for increased Ca levels in seed also may be beneficial in overcoming seedling diseases. Many of the Ca deficiency problems that occur during germination probably develop because of the inherently very low concentrations of Ca in cottonseed (46,82).

The effects of other nutrient elements on seedling diseases are not as well-defined as those of Ca. Omitting K, P, or S from nutrient solutions increases the incidence of damping-off from *R. solani,* and the occurrence of damping -off in both sand-culture and nutrient-depleted soils is inversely related to K concentrations (119). Fertilization with N increases *R. solani* infection at normal planting temperatures, 21-24 C, but decreases infection at 33 C (119,179). Fertilization with asparagine breaks down the resistance of *Gossypium arboreum* L. 'Nanking' seedlings to *Colletotrichum gossypii* Southworth; inorganic sources of N, except for $(NH_4)_2SO_4$ which is phytotoxic, do not show this effect (21). Other amino acids are much less effective than asparagine (21). This may indicate that generation of free NH_3 by the pathogen may be involved in the phenomenon, because asparagine is an excellent substrate for generation of NH_3 by microorganisms (50,57,79). Fertilization with P, or inoculation with mycorrhizal fungi that increase P uptake, generally does not decrease the frequency of seedling diseases, but may facilitate recovery from debilitating infections by *Thielaviopsis basicola* (138,179). Soaking seed in 0.1% $MnSO_4$ may increase the rate and percentage of germination, and decrease the percentage of diseased seedlings by as much as 25% in the field (63). Other microelements also stimulate cottonseed germination in laboratory tests, but these elements have not been beneficial in the field.

Fusarium wilt

Fusarium wilt occurs throughout the cotton-growing areas of the world and is caused by eight physiological races of *Fusarium oxysporum* Schlect. f. sp. *vasinfectum* (Atk.) Snyd. & Hans. Races 1, 2, and 6 attack Upland cotton (*G. hirsutum*) and Egyptian cotton (*G. barbadense*), but are not virulent to the Asiatic diploid cottons (*G. arboreum* and *Gossypium herbaceum* L.). The most severe diseases caused by races 1, 2, and 6 occur in sandy acidic soils containing appreciable populations of root-knot nematodes in areas of moderate to high rainfall. Race 1 is the most common and occurs in the Americas, Africa, Europe, and Asia, including the parts of the USSR and India where *G. hirsutum* is grown. Only races 1 and 2 have been isolated from cotton in the USA.

Races 3 and 5 of *F. oxysporum* f. sp. *vasinfectum* attack Egyptian and Asiatic cottons, but not Upland cotton, and have long been a problem in the Nile Valley and other areas of Africa where *G. barbadense* is grown. Race 4 attacks only the Asiatic species and occurs in India and other Asian countries where *G. arboreum* and *G. herbaceum* are grown. Races 3, 4, and 5 are favored by heavy (clay) soils with alkaline pH values of 8.0-8.3 (32,134). These races often occur in association with the reniform nematode (33), which is better suited to the heavy soils than the root-knot nematode. Races 7 and 8 of *F. oxysporum* f. sp. *vasinfectum* have been reported recently from China.

Predisposition by nematodes is essential for severe infections by races 1 and 2 in the field. Consequently, disease caused by *F. oxysporum* in the USA is often referred to as the Fusarium wilt-nematode complex. This fact is extremely important in understanding the effects of macro-

and microelements on the disease; elements may affect the fungus (*Fusarium*), the nematode, the fungus-nematode interaction, or the resistance of the host to either pathogen.

The association of Fusarium wilt (races 1 and 2) with acidic soils was clearly shown in a survey study of fields in 14 counties in Texas (161). Fusarium wilt occurred in 55% of the fields with acidic soils (pH 5.5-6.4), in 13% of the fields with neutral soils (pH 6.5-7.4), and in only 2% of the fields with alkaline soils (pH 7.5 and above). When viewed in terms of basicity, the disease occurred in 32% of the fields with low basicity (0.0-0.9%), in 8% of those with medium basicity (1.0-2.5%), and in no field with basic soils (2.5-10.0%). Albert (4) used nutrient-sand cultures to show that wilt developed more rapidly and extensively at pH 5.0-5.3 than at pH 6.7-6.8 regardless of N source. Fusarium wilt caused by race 1 or 2 has been observed on a few occasions in sand or loam soils at pH levels as high as 8.1-8.2 (19). In these cases the disease was associated with high populations of nematodes and saline conditions, and there was no indication that wilt was extensive or severe.

The possible beneficial or adverse effects of inorganic fertilizers on Fusarium wilt were debated as early as the beginning of the twentieth century. However, early experiments on soils of the Southeastern Coastal Plain of the USA failed to show clearly either beneficial or adverse effects of N, P, or K (71,109). The failure to demonstrate effects of elemental amendments probably was due to very high populations of root-knot nematodes, the use of very susceptible cultivars, and the very high inoculum densities of *Fusarium*. Where these conditions exist, high percentages (60-100%) of plants are killed, and macroelement amendments seldom show more than minor effects (88,113,163,164,166). The macroelements also show only minor effects, if soils contain plentiful K (32,88).

In 1924 Rast (123) first reported remarkable control of Fusarium wilt in Arkansas with K fertilizers. He also noted that N and P without K did not reduce wilt frequency. Subsequent studies by V. H. Young, Ware and colleagues in Arkansas (167,177,178), Neal and Miles in Mississippi (100,101), Tisdale and Dick in Alabama (163,164), and P. A. Young (175) in East Texas clearly showed that K deficiency predisposes cotton to Fusarium wilt. Various sources of K, and applications either before or after planting, were equally effective in reducing the percentages of Fusarium wilt on K-deficient soils, especially if only low nematode populations were present. These studies also showed that N or P fertilizers generally aggravated existing K deficiencies and increased the severity of wilt. Nevertheless, combinations of N and P often increased yields

172

substantially. Combining K with N and P in balanced fertilizers proved to be the best treatment to decrease wilt and obtain maximum yields.

Increases in wilt severity with N fertilization, and decreases with K fertilization, also have been reported for neutral and alkaline soils of Africa and India, which are infested with races 3, 4, or 5 (32, 34,96,134). At these locations P deficiency has been reported to increase incidence of disease, and fertilization with P decreases incidence of disease. In Brazil macroelements affected wilt in the same manner (99). Fertilization with Zn also has given decreased percentages of wilt in Indian soils, while other microelements have given variable results (134).

The effects of K, N, and P on wilt severity also have been studied in sand-nutrient cultures with race 1 (or 2) in the USA (1,4,100) and with race 3 in Egypt (98,142,143,144,145,146,147). With race 1, the least disease occurred with low P and N levels; high N levels increased disease severity, and high K levels decreased onset and severity of disease. With race 3, low concentrations of N, K, or P, compared to nutrient solutions (pH 6.8) completely deficient for each one, reduced disease severity in a susceptible cultivar (147). As the concentration of each element was increased, disease severity increased to a maximum, and then declined with further increases in concentrations. Maximum wilt (100%) occurred with 200 mg/L of K (145), 200-300 mg/L of P (98) and 300 mg/L of N (98) in the nutrient solution. Increasing K levels to 1000 mg/L reduced wilt to 11%, while the same concentration of N decreased wilt to 36%. The highest P concentration used (500 mg/L) only reduced wilt to 90% compared to 100% for the control. Nutrient deficiencies or excesses did not change the reaction of the resistant cultivar to the pathogen. In the susceptible cotton cultivar, inoculation with the pathogen generally increased the uptake of P, Ca, and Mg, but decreased the uptake of K and N by the plant (143,144,146). High levels of N also caused a decrease in K uptake, while increasing Ca and Mg uptake (146).

The effects of N sources on wilt severity also have been investigated. At low pH (5.0-5.2) in sand-nutrient cultures, (NO_3^-) N gave more rapid disease development than (NH_4^+) N; whereas at neutral pH (6.7-6.8) (NH_4^+) N gave the most severe disease (4). The differential effects of N sources on disease severity at the two pH levels are partially due to maintenance of favorable pH for disease development; (NO_3^-) cause increases in pH and (NH_4^+) salts cause decreases in pH during plant growth. In pure cultures *F. oxysporum* grows and sporulates more readily on (NO_3^-) than (NH_4^+) N sources (67,97). In the field percentages of wilt obtained with different N sources generally have not been significantly different, and all sources have been observed to increase disease severity when K is deficient.

The impact of K deficiency and fertilization on Fusarium wilt depends both on the level of resistance in the cultivar and on the population density of accompanying nematodes (33,43,88,163,164). In a controlled nutrient study, a resistant cultivar of Egyptian cotton (*G. barbadense*) showed 12.5% wilt with low K (50 mg/L), whereas wilt was completely suppressed at 100 mg/L of K, if only the fungus were present. When both the fungus and reniform nematode were present, 500 mg/L of K were required to completely suppress disease. In a susceptible cultivar significant reduction of disease occurred only at 500 mg/L of K with the fungus alone, and at 1000 mg/L with both of the pathogens. Concentrations of K in the resistant cultivar were 1.42% and 2.38% when complete resistance was achieved to the fungus alone and to both pathogens, respectively. In susceptible plants a concentration of 3.13% K was required for increased tolerance even to the fungus alone. Similarly, in Alabama fields 8% K in mixed fertilizers was required to give maximum suppression of wilt in a susceptible cultivar; whereas only 4% K gave maximum benefits with a resistant cultivar. Thus, the dose of K required to suppress wilt decreases as cultivar resistance to *Fusarium* or nematodes increases. The required dose increases as the population density of the nematode increases. No benefits are obtained from K with very high nematode populations and susceptible cultivars (88,163). The reduced responsiveness of wilt to K in the presence of nematodes may be due to the greatly reduced concentrations of K in roots and shoots, that are caused by nematodes alone (111).

Various attempts have been made to explain the effects of elements on Fusarium wilt. Elements may affect the survival and saprophytic growth of *Fusarium* in the soil, the virulence or pathogenicity of the fungus, or the resistance of the cotton plant to the disease.

The *Fusarium* pathogen requires N, P,. S, Mg, and K as macroelements (97), and Zn, Mn, and Mo as microelements for optimum growth and sporulation in pure culture (64,134,174). Other microelements (Fe, Cu) undoubtedly are required, but occur in sufficient quantities as contaminants in the major salts used in nutrient solutions. Nitrate, (NH^{+}_{4}), or organic sources of N adequately support growth; but best growth is obtained with (NO^{-}_{3}) N. Glucose and $C_{12}H_{22}O_{11}$ are the best C sources for growth and sporulation.

Ammonia, but not (NH^{+}_{4}) ion, is toxic to *Fusarium* species, which may account for the suppression of disease by high concentrations of certain N fertilizers when optimum concentrations of other elements are present. Fusarium propagules were killed up to 4cm from a point source of NH_3, and did not recolonize the treated soils for a period of 225 days (149). Part of the toxicity from NH_3 in natural soils apparently is due to nitrite ions formed by the biological oxidation of NH_3. more than 30-35 and 100 mg/kg of nitrite accumulated at 24 and 6 C, respectively, in a silt loam soil by 4 days after introduction of NH_3 (149). Adding 35 mg/kg of nitrite to soil gave complete kill of *Fusarium* populations. Sensitivity to NH_3 might be involved in suppression of races 1, 2, and 6 by neutral or alkaline soils and in suppression by biological antagonists. Various microelement amendments also inhibit conidial germination and sporulation of *Fusarium* when added to soils at 50-400 mg/kg (134), but these concentrations may be rare in natural soils.

Suppression of wilt by N and P in certain situations may involve affects on natural soil microflora, which in turn affect populations of *Fusarium*. Nitrogen fertilizers generally cause increases in bacteria and fungi, including *Fusarium*, in the rhizosphere (26,134). Fertilization with K may greatly increase percentages of fusarial antagonists in these population (134). Thus, fusarial populations in the rhizosphere may increase in response to N when K concentrations are low; but when K is high, and increase of antagonists may prevent the increase of *Fusarium* populations. High concentrations of N decrease K levels in the plant and might have a similar effect of P are on beneficial microorganisms (99).

Inoculum from element-deficient media is less virulent than that from nutrient-complete media (134). Consequently, various attempts have been made to explain the effects of elements on wilt severity by their effects on virulence and pathogenicity. The *Fusarium* pathogen produces fusaric acid, which is extremely toxic to cotton plants, and has been demonstrated in the diseased plant. Pectinase enzymes are also thought to be involved in tissue degradation. Critical concentrations of Zn are essential for production of both fusaric acid (64) and pectin methyl esterase (PME) (134). Iron also promotes PME production (134). In contrast, K inhibits fusaric acid production, as well as fungal growth, in melons (120). Fusaric acid production and sporulation on synthetic media also are much greater with $NaNO_3$ than with K or $Ca(NO_3)_2$ (67) or other N sources (24). Thus, acidic sandy soils may provide the relatively high Fe and Zn and low K and Ca concentrations needed to maximize production of fusaric acid and PME.

Several observations indicate that the major effects of K are on host-defense responses (134,135). The effects of K are equal in sterile and nonsterile soil, indicating that the primary effects are not on the soil microflora. Also, increased genetic levels of resistance in the host diminish the amounts of K required to enhance resistance. The Resistance of cotton to Fusarium wilt has been correlated with concentrations of gossypol in root tips, toxicity of root extracts to *F. oxysporum*, speed of accumulation of terpenoid aldehyde phytoalexins and tyloses in xylem vessels, and sensitivity to phytotoxins produced by *F. oxysporum* f. sp. *vasinfectum* (15). Unfortunately, the effects of elements on these characteristics are unknown.

Verticillium wilt

Verticillium wilt of cotton is caused by *Verticillium dahliae* Kleb., which includes 16 known vegetative compatibility (v-c) groups (118). Most isolates from cotton in the USA belong to v-c group 1; isolates of v-c group 2 also have been obtained from California and Texas. Isolates belonging to v-c groups 3, 6, 7, and 8 have been isolated from cotton in other parts of the world. Isolates in v-c group 1 cause a severe defoliating-type of disease and are more virulent to cotton than isolates from other v-c groups. Currently, v-c group 1 is known to occur only in Peru, Mexico, and the USA.

V. dahliae survives in the soil as dormant microsclerotia that may remain viable for many years. Root exudates stimulate germination of the microsclerotia, and the fungus subsequently penetrates the root to enter the vascular system, where it spreads throughout the plant. Nematodes are not essential for high levels of infection, but when present in large numbers, they may increase the incidence of wilt (89). Disease frequency and intensity are usually directly related to the concentrations of microsclerotia in soils. Thus, elements may alter disease by changing densities of microsclerotia or levels of host resistance.

Verticillium wilt of cotton is most prevalent in neutral to alkaline loam and clay soils in areas where there is limited rainfall, and crops normally are irrigated. The disease shows an extremely strong response to temperature and normally appears in the field during the later half of the growing season, when mean temperatures fall below 27 C (14). Irrigation that lowers soil temperature often triggers the first appearance of the disease.

The general absence of Verticillium wilt from areas with acidic soils may be due to the toxicity of Mn, Al., and possibly other acid-soluble microelements. When Mn-toxic soil is limed to increase the pH from 4.6 to 6.3, susceptible, tolerant, and resistant cultivars of cotton show increased severity of wilt following inoculation by increased severity of wilt following inoculation by stem-injection with isolates from wither the v-c 1 or 2 groups (141). Such liming treatments also greatly reduce Mn concentrations in plant tissues (41). Similarly, liming Al-toxic Bladen soil to increase the pH from 4.4 to 5.4 greatly increases the severity of Verticillium wilt in sunflower (108), and decreased Al concentrations in cotton plants (153). Treatment of naturally-infested soils in South Africa with Al SO_4, to decrease pH from 8.0 to 4.5, decreased Verticillium wilt of cotton from 55% to 0%, and greatly reduced survival of the MS in soil(8). Amendments with S (2,000 lb/A), in addition to 600 lb/A of N-P-K (6-8-12) fertilizer, reduced the mean percentage of wilt in Arkansas from 32.5% to 24.4% during a three-year period, while limestone (2,000 lb/A) increased wilt incidence to 38.2% (176). Amendments with S did not control wilt in California (130), or reduce wilt in New Mexico (47). However, in the latter case, only 500 lb/A of S was applied to highly calcareous soils (pH 8.4-8.5), and it is unlikely that the treatment appreciably changed pH.

Availability of K is critical for the resistance of cotton to Verticillium wilt, just as it is for resistance to Fusarium wilt. There are many reports of K fertilization reducing wilt severity (1, 5, 45, 55,114, 115, 176). Presley and Dick (115) showed that although K fertilization was beneficial in K-deficient soils in Mississippi, it did not reduce wilt severity in soils naturally high in K. This probably accounts for the lack of response of wilt to K in the Southwestern USA (47), where soils have adequate K supplies. Even in deficient soils, the effectiveness of K in reducing wilt depends on level of host resistance and density of microsclerotia in soils. As with Fusarium wilt, Verticillium wilt is most effectively reduced when K is used with resistant cultivars, and in soils with low or moderate infestation of the pathogen (5).

Considerable evidence indicates that *V. dahliae,* and perhaps other root pathogens, aggravate K deficiency in cotton by further reducing K levels in tissue. About 300,000 acres (15-20% of the fields) in California are affected by a K deficiency-disease complex that frequently, but not always, involves *V. dahliae* (5,45,170,171). Leaves of plants affected by the complex show a bronzing and metallic sheen that is typical of K deficiency; and marginal or deficient levels of K also are present in petioles and leaves. Yet, soil analyses indicate that many of the fields have adequate levels of K, and corn grown in soil from these fields does not show K deficiency (5). Where *Verticillium* is present with the complex, K fertilization usually decreases the percentage of Verticillium wilt and gives small yield increases, but it does not alleviate the deficiency symptoms. Soil fumigation (45) or soil solarization (115,170) greatly reduces Verticillium wilt, eliminates K-deficiency symptoms, and gives pronounced yield increases, but only slightly increases the available K in soils. Collectively, these observations indicate that *V. dahliae,* and probably other root pathogens, are partially responsible for the K deficiency in plant tissues. Recent unpublished results with deep chisel placement of K in the subsoil, as well as in the plow layer, indicate that K deficiency in the subsoil also may contribute to the problem.

The K-binding clays, such as those in the San Joaquin Valley of California, possibly create conditions for deficiency-disease complexes. Leaf bronzing and weak response to K fertilizers also have been observed in Central Texas and Mississippi in recent years. Thus, K deficiency-disease complexes on K-binding soils may be more common than previously realized. This possibility can only be resolved by careful analyses of K and Na levels in petioles and leaves, and by better definition of what constitutes critical K concentrations of different types of soils and with different concentrations of Na, which may partially substitute for K.

In sand-nutrient cultures, both high and low concentrations of (NH^+_4) and NH_3 N may give less wilt than intermediate concentrations (121). However, in the field heavy N fertilization generally increases the apparent severity of Verticillium wilt, especially when plants are deficient in K (25,73,114,115,176). In the absence of other elemental deficiencies, low or moderate amount of N fertilization may not significantly increase wilt percentages (8,35,47). Even when wilt percentages are increased, N fertilization generally gives yield increases that are proportional to the N doses. Thus, it is often not economical to control wilt by using low levels of N fertilizer. Using balanced fertilizers that correct K and P deficiencies usually limits the increase in wilt percentages caused by N and maximizes yields. Split applications of N fertilizers (36,37,87) or applications of N as a side-dressing midway through the growing season (104), may further reduce wilt and increase yields.

All N sources have been reported to increase percentages of wilt in certain situations. However, high concentrations of urea increase the disease less than other N sources in sand nutrient culture (121). When applied several weeks before planting, or as 2% sprays to foliage, urea also has decreased wilt percentages in; the field (25,55,90). At certain concentrations, NH_3 N also gives significantly (121) and in the field (25). Alfalfa hay, an organic source of N, combined with inorganic N, gave the least wilt and greatest yield in Arkansas (176). Manure, on the other hand, generally increased Verticillium wilt more than balanced inorganic fertilizers (47,176).

In the greenhouse, fertilization of P-deficient soil with superphosphate (calcium dihydrogen phosphate) increased the severity of Verticillium wilt caused by defoliating (v-c 1) isolates (30). Inoculation of the soil with *Glomus fasciculatus* (Thaxt. sensu Gerd.) Gerd. & Trappe nov. comb., a mycorrhizal fungus that increases P uptake from deficient soils, also increased severity of wilt. Plants with more severe symptoms contain greater numbers of propagules in perioles, indication that P favors development of the fungus within the plant. Under field conditions, superphosphate fertilization alone did not change the severity of wilt (47,176), but deficiency conditions may not have existed.

Phosphorus, combined with very high rates of N, has increased wilt severity more than N alone (73). Combinations of P with lower N rates also increased wilt when K was deficient (115), but not when K was adequate (8). Thus, the effects of P depend on the concentrations of other elements, as well as whether P deficiency exists, and generally are small in magnitude.

Microelements may either suppress Verticillium wilt or have no effect, depending on concentrations in the growth medium and in the fertilizer. Benefits apparently are obtained from microelement fertilizers only when deficiencies exist. No changes in wilt incidence or yield were obtained from complete mixtures of microelements in Arkansas (176) or Texas (73), where microelement deficiencies are unknown. Under greenhouse conditions, B and Mn gave slight reduction of wilt in Mississippi soils (31), and sprays with $CuSo_4$ before inoculation decreased wilt severity in California (6). However, in neither case were benefits shown in the field. Relatively high concentrations of Mn (141) and Al (108) in soil, Zn in sand-nutrient culture (59), or Cu in nutrient solutions (13) markedly suppressed wilt, but also were phytotoxic. Thus, these treatments are of little practical use.

In Russia, microelement mixtures and various individual microelements frequently have decreased wilt severity and given appreciable yield increases (9,10,55,56,86,90,95,137). Microelements may decrease wilt severity in Russia because deficiencies that aggravate wilt are more common than in the USA. Beneficial effects have most commonly been obtained with Zn, Mo, Bo, and Mn. Sprays applied at budding, flowering and during fruit development have been used most commonly. However, decreased disease severity, and increased yield, also have been obtained with soil applications before planting and from seed treatment (10,55). Combining Zn with 2% urea in sprays has proven useful for providing both Zn and N (90). As cotton production continues in the USA, microelement deficiencies undoubtedly will become more prevalent, and may contribute to wilt severity if left uncorrected.

Salinity apparently affects the distribution and severity of wilt in the Southwestern USA. In certain locations in Arizona the severity of wilt in cotton fields is negatively correlated with the electrical conductance of saturation extracts from soils (28). The incidence of wilt is significantly less where conductance exceeds 5 millimhos per cm; above 10 millimhos little wilt occurs. Similar results are obtained when these conditions are duplicated under greenhouse conditions. In cotton-leaf samples from the field, Na concentration in disease-free plants was about four times greater than that in nearby plants that were severely affected with wilt. Thus, Na, as well as K, apparently can enhance cotton resistance to wilt. High plant-beds that increase salt concentrations in the root zone also reduce wilt in saline soils, whereas excessive irrigation lowers salt concentrations in the soil solution and increases disease severity.

Efforts to explain the effects of elements on Verticillium wilt severity have centered on the growth and survival of *V. dahliae*. The fungus grows best at pH 6.0 to 7.0, and uses (NH^+_4) ions, (NO^-_3) ions, or urea as sole N sources in pure culture (81,139,162). Best growth, especially conidiation, is obtained when N is supplied as amino acids (particularly asparagine, glutamine, alanine, arginine). Malca *et al.* (81) concluded that (NH^+_4) ions were superior to (NO^-_3) ions as an inorganic N source, but Selvarj (139) reached the opposite conclusion. *Verticillium dahliae* requires supplemental Zn, Fe, and Mn, as well as N, P, K, S, and Mg for its optimal growth in pure cultures (92,162). Deficiency of microelements decreases the rate and percentage of germination of conidia, especially with low water-activity (92). Deficiency of Zn causes decreased N concentrations in the fungus, and responses to Zn vary with different N sources (162). Requirements for other microelements have not been established.

The suppression of Verticillium wilt in acid soils has been attributed to the toxicity of Mn and Al ions. Both of these elements inhibit growth and microsclerotia production in acid (pH 4-5) cultures. Mn inhibits extension of mycelia at 120 mg/L and production of microsclerotia and accumulation of dry weight at 30 mg/L (141). Al at 8 mg/L suppresses growth and almost completely prevents MS formation (108). Adjusting soils to pH 4.5 with $Al_2(SO_4)_3$ reduces viable propagules by 90% or more (8). Moisture and N and P fertilization enhance the toxic effects of the $Al_2(SO_4)_3$. Mixtures of microelements that reduce wilt in the field also inhibit growth of *V. dahliae* in culture (10), but it has not been determined whether concentrations similar to those in culture also occur in soil solution or plant tissue.

Reduction of wilt by high concentrations of urea or (NH^+_4) fertilizers in certain situations also may be due to repressed growth and survival of the fungus in the soil. Reduced numbers of propagules in soil have been noted following the application of $Al_2(SO_4)_3$ (89,173),

$NH_4H_2PO_4$ (25,154), NH_4NO_3 (8), and urea (25,44). Ammonium fertilizers also enhance killing of the pathogen by soil solarization (154). In contrast to (NH^+_4) fertilizers, (NO^-_3) fertilizers increase numbers of *V. dahliae* propagules in soil (25). It has not been determined whether free NH_3 generated from(NH^+_4) or urea fertilizers may kill microsclerotia of *V. dahliae*.

The (NH^+_4) N stimulates more biological activity than (NO^-_3) N in the soil (25), and might favor the build up of antagonists. In the only specific case studied, N (unspecified source) increased total populations of fungi in soils, but greatly decreased the percentages of antagonists to *V. dahliae* (65).

The effects of elements on virulence are largely unknown. Calcium (0.005-0.05 M) stimulates production of polygalacturonase and pectin transeliminase in pure culture, and both Ca and Na ions up to 0.05 M progressively increase the activity of the enzymes (140). Calcium also increases the toxicity of purified polygalacturonase from *V. dahliae* to cotton cuttings (94). These effects suggest that Ca amendments may increase disease severity by increasing virulence. However, experiments with enzyme-deficient mutants (49) indicate that pectinase enzymes are not important for virulence.

Phytotoxic concentrations of Cu ions in nutrient solutions elicit phytoalexin synthesis in paravascular cells of cotton xylem vessels (12; Bell, unpublished). Certain fertilizers containing (NH^+_4) ions also have caused massive production of phytoalexins in stele tissues at toxic concentrations (23; Wilhelm personal communication). Since *V. dahliae* is inhibited by these phytoalexins at 25-100 ppm (80), decreases in Verticillium wilt caused by phytotoxic concentrations of elements may involve induced resistance reactions in the host.

Phymatotrichum Root Rot

The fungus *Phymatotrichum omnivorum* (Shear) Dug. is widely distributed in soils of the Southwestern USA and Northern Mexico where it causes devastating diseases of cotton and many other dicotyledonous plants (77,156). It does not occur anywhere else in the world. Because of the impact of the disease on cotton, and its prevalence in Texas, it also is referred to in the literature as cotton root rot and Texas root rot. The disease has long intrigued microbiologists and ecologists because it occurs in well defined areas of fields year-after-year with only limited spread into noninfested areas (77). Certain areas of the fields apparently remain uninvaded indefinitely, and other areas, once invaded, may "recover" and remain uninfested for several years. Thus, soil conditions appear to have strong effects on the survival and spread of the fungus in soils.

Phymatotrichum omnivorum survives as free sclerotia in soil or as mycelial strands on live roots. The disease is spread by growth of fungal strands through the soil from infected plants to adjoining healthy plants, where infection is repeated. When sufficient C source is available, the strands differentiate into sclerotia. Spore mats occasionally form on the soil surface, but the resulting spores do not readily germinate and apparently are not involved in the spread of the pathogen. Because the sclerotia are formed and survive at depths as great as 1 m in soil, it is very difficult to eradicate the fungus from soils.

In the late 1920's and the 1930's, extensive studies of the ecology of *P. omnivorum* were conducted in Texas, especially at the Greenville Station. These and later ecological studies have been reviewed in detail (77,156). A survey reported in 1928 (161) first established that the disease in Central and Eastern Texas was associated mostly with alkaline soils. In fields with a pH of 5.5-6.4 the disease occurred only in trace amounts and in only 34% of the fields. In contrast, 10-100% of the plants were killed by root rot in 60% of the fields with a pH of 6.5-7.4, and 71% of the fields with a pH of 7.5-9.0. Subsequent studies (38,39,159,160) clearly demonstrated the strong effects of pH on the pathogen and the disease. In inoculated soils, there is a marked diminution of root rot at pH 6.0, and virtually no root rot at pH 5.0. At pH 6.0-7.0, high percentages of infection occur following inoculation, but compared to higher pH's, there is a progressive decrease in the ability of the fungus to spread from plant to plant or survive from 7.0 to 6.0. The greatest infection, spread, and survival by the pathogen occurs at pH 7.5-8.5. Comparisons of areas with and without root rot within the same field showed that areas with root rot had mean pH values of 7.4-8.3 compared to 5.7-6.3 in areas where no disease occurred (38). These early studies further showed that alkaline conditions resulted from high concentrations of Ca carbonates, and soils favorable for disease generally contained in excess of 30% clay (25,152). The disease was most severe in the calcareous Houston black clay of the Blackland prairies of Central Texas, but also occurred in at least 30 other soil types. Disease-favorable soils have since been characterized as Vertisols, i.e., soils dominated by shrink-swell clays, crack formation to depths of 0.5 meter, and cation exchange capacity generally above 35 cmol/kg of soil; the clays are montmorillonite (152).

Higher concentrations of total and available N, P, K, Mg, and Ca occur in the root rot-conducive soils of the Texas Blackland prairies than in those of the root rot-suppressive fine sandy loams of East and Central Texas (42). Thus, deficiencies alone (at least) cannot account for disease severity. Comparisons of elements in soils from diseased and nondiseased areas of the same field have indicated some differences in elemental concentrations that might contribute to disease incidence. Lyda (74,75) found that areas with root rot-diseased plants generally contain lower concentrations of Na than areas with plants free of root rot, and he used NaCl amendments to reduce disease incidence in the field. Subsequent, more extensive, surveys failed to confirm the association of high Na levels with disease suppression in either Texas (152) or Arizona (93). Diseased portions of fields in Texas generally have lower Mg concentrations and higher total Ca/Mg ratios, soluble Ca concentrations, and soluble bicarbonate ion concentrations, that areas free of disease (152). In some areas low Fe and Zn concentrations also are associated with conducive soils. The association of root rot with lower Mg concentrations also had been observed in a few fields in Arizona (93). Thus, nutritional stress may be associated with areas of fields conducive to root rot. It is essential that foliar analyses be used to confirm that the differences observed in soils actually contribute to deficiencies in plant tissues.

Based on the ecological studies, attempts have been made to control root rot by adjusting pH downward to pH 5.5, or below, with S. In noncalcareous soils the percentage of infected plants has been reduced, and yields have been increased significantly (39,160). Highly calcareous soils, however, have so much buffering capacity, that it is not economically feasible to change pH with S. In addition, high rates of S application are phytotoxic to cotton (160). Thus, S fertilization is useful in only a very small percentage of fields affected by root rot.

Soils infested with *P. omnivorum* generally are responsive to N fertilization, and many such soils are slightly deficient in P. Both of these elements have been studied extensively for their effects on root rot. Nitrogen from mixed sources used alone, or as the predominant macroelement, reduces percentages of Phymatotrichum root rot and Macrophomina root rot (Abyan root rot); whereas P alone, or as the predominant macroelement increases Phymatotrichum root rot in various soils (3,18,60,61,72,128,155). When these two elements are used in similar amounts, there is no effect on root rot percentages (124). In all cases the greatest yield increases and economic returns result when recommended amounts of P are used with N.

Sources of N also have been examined for their effects on control of Phymatotrichum root rot. Manure, buried under the planting row, has given remarkable control of root rot in Arizona (69,155). Urea, used as the N source, also has reduced percentages of root rot; whereas Ca $(NO_3)_2$ has aggravated root rot in some experiments (85,91). During 1977, a dry year with low incidence of root rot, deep chisel applications of anhydrous NH_3, alone or with N-Serve or azide, significantly reduced percentages of plants dying with root rot in Blackland prairies of Texas (131). However, over a five-year period, the NH_3 treatments were no better than the deep-chisel control (133). Similar inconsistent results with anhydrous NH_3, NH_4OH solutions, and $(NH_4)_2SO_4$ have been obtained on the Coastal Plain prairies of Texas (83,84,85,91) and in Arizona (68,155). A heavy application of $(NH_4)_2SO_4$ (4,350 lb/A), or the same rate of $NH_4H_2PO_4$ (watered with 3 to 4 inches of irrigation water) eradicated *P. omnivorum* from certain trees (155), but these doses are toxic to cotton (68). Recent results (91) indicate that repeated use of $(NH_4)_2SO_4$ as a fertilizer at recommended rates over several years may have cumulative benefits in controlling root rot, but this possibility needs further testing.

Macroelements other than N and P have not been shown to affect root rot. Soils infested by *P. omnivorum* generally have excess Ca and are sufficient for K. Thus, it is not surprising that these elements have no effect. Although lower Mg concentrations recently were associated with the disease, no concerted effort has been made to determine the effects of Mg fertilization on the disease. This needs to be a priority subject for future research.

Sodium chloride at 900 and 2400 lbs/A reduced the percentage of plants killed by root rot in the Blackland prairies of Texas (76,78). The suppressive effects persisted for 2 years with the lower rate, and strong suppression was maintained for 3 years with the higher rate. Amendments with Na have given variable results in the Coastal Plain prairies of Texas (84), and do not reduce root rot in Arizona (93). Treatment of Houston black clay with NaCl to give 50% saturation increased the pH from 8.2 to 9.53 (78), which may be adverse for cotton yields. This level of Na also aggravates certain deficiencies, e.g., those of Mg, Fe, and Zn, that may exist in root rot-conducive soils.

Soils of the Southwestern USA generally are not considered to be deficient in minor elements, with only a few local exceptions for Zn, Fe, Mn, and Cu. Surveys of root rot areas of fields indicate that localized deficiencies of Fe and Zn, and possibly Cu, are associated with root rot severity. Accordingly, treatments with Fe and Zn, particularly as drenches around the stem, have slightly reduced root rot percentages in some years and locations (85,91).

Most efforts to explain the effects of pH and of elements on root rot have centered on the growth and survival of the fungus in soil. Lyda (75,76) proposed that the association of root rot and *P. omnivorum* with calcareous alkaline clay soils is due to beneficial effects of CO_2 on growth and development of the fungus. Summer rains that trigger germination of sclerotia and infection of the plant also cause marked build-up of CO_2 concentrations in Vertisols. Treatments of moist calcareous soils with 0.5-5.0% CO_2 in air stimulates the germination of sclerotia and strand growth. In the presence of substantial organic matter, production of sclerotia from the strands also is stimulated.

The active stimulatory molecule in these cases apparently is the bicarbonate ion, which is formed from CO_2 under the alkaline conditions in the soil solution (76). In acid soils the CO_2 is converted primarily to carbonic acid, which is not beneficial to the fungus. Likewise, the high pH created by Na treatments causes the CO_2 to be converted mostly to carbonate ions, which may be toxic to the fungus. In Houston black clay, 50% and 100% saturation with Na reduced total sclerotial production from 6.52 to 0.63 and 0.12 g per culture, respectively. The mean dry weight of individual sclerotia also was reduced from 1.64 mg to 0.30 mg by the complete saturation with Na.

Attempts to explain the effects of N on root rot have centered on the nutritional requirements of the fungus and the possible toxic effects of NH_3 and (NH^+_4) ions. In 1932, Neal (105) first reported that $P.$ $omnivorum$ is extremely sensitive to NH_3. Adding NH_3 to cultures, to give 0.1% in the medium, killed the fungus within 20 minutes; exposure to NH_3 gas for 30 seconds killed fungal strands and sclerotia, even on infected roots. Growth of the fungus in Duggar's solution for fungi was markedly restricted also by NH_4NO_3 and $(NH_4)_2SO_4$, whereas Ca, Na, and K nitrates produced abundant growth (106). Thus, NH_3 and(NH^+_4) fertilizers were suggested as possible treatments to control root rot in the field.

Subsequent detailed studies (17,20,40,157,158) of the nutrition of $P.$ $omnivorum$ established that (NH^+_4) ions, in fact, were not toxic to the fungus and could be used readily as the sole N source. The restriction of growth by (NH^+_4) ions in Duggar's solution is due to the low pH which develops when (NH^+_4) ions are used as the only N source (20); growth is completely inhibited at pH 3.0 (40). At a amino acids, peptone, urea, (NH^+_4) ions, (NO^-_3) ions or nitrite ions as a sole N source (40,158). In addition to N, $P.$ $omnivorum$ requires the macroelements P, K, Mg, and S (40,157), and microelements Fe, Zn, and Am (17), for optimal growth. Slight growth increases are also obtained with Cu and B, whereas 4 mg/L Ni and Co are toxic (17). With adequate glucose growth rate is proportional to N supply, and is best with 2 mg/L of each required minor element (17,157). Growth with (NO^-_3) ions is influenced by the balance between K partially substitute for K and Mg, respectively (157,158). Most elements are tolerated in a rather broad range of concentrations, when proper balances are maintained.

The toxicity of NH_3 to $P.$ $omnivorum$ has been confirmed (102,103,132) and toxic thresholds per volume of tissue have been established (132). Mycelia are killed by exposure to 4 mg/ml of NH_3

for 1 min, and strands in infected roots are killed by 56 mg/ml within 24 hours. Toxicity to sclerotia at 28 mg/ml is proportional to exposure time; 3, 23, 34, and 59% of sclerotia were killed after 1, 12, 24, and 48 hours. Exposure to 42 mg/ml for 12 hours completely killed sclerotia.

Much higher concentrations of NH_3 are needed to kill sclerotia in soil, probably because of the conversion of NH_3 to (NH^+_4) and (NO^-_3) ions. In soil with 27-30% moisture, 1,025 mg/ml of NH_3 were required to prevent growth of *P. omnivorum* (103). Similarly, 138 and 276 mg/ml were required to obtain 35 and 79% kill of sclerotia 9132). The NH_3 causes electrolyte leakage and marked inhibition of respiration in the fungal cells.

The degree to which NH_3 generation is involved in the suppression of *P. omnivorum* by N fertilizers is uncertain All of the inorganic (NH^+_4) N sources should give some NH_3 in alkaline soils. Urea also decomposes to NH_3 in alkaline soils. The effectiveness of manures is associated with microbiological activity (29,69), and might involve the ammonification of the organic N in the manure. If NH_3 is involved in the suppression of *P. omnivorum* , $(NH_4)_2PO_4$ might be the most useful preplant fertilizer because of its ability to generate NH_3 in soils.   Likewise, inoculating cotton stalks with NH_3-generating bacteria, such as *Enterobacter cloacae* (50,51), before plowdown, might increase eradication of *P. omnivorum* in soils.

Nematodes

The most important nematodes attacking cotton are the sedentary endoparasites *Meloidogyne incognita* (root-knot nematode) and *Rotylenchus reniformis* (reniform nematode) (165). *Meloidogyne incognita* is by far the most important nematode pest of cotton on sandy soils throughout the world, whereas *R. reniformis* is often an important problem on loam or clay-loam soils with a neutral to alkaline pH, especially in Africa, Asia, and in Southern Texas and Louisiana. The migratory nematode pests of cotton include the lance nematodes (*Hoplolaimus columbus* Sher. and *Hoplolaimus galeatus* (Cobbs) Thorne), the sting nematode (*Belonolaimus longicadatus* Rau), and the stunt nematode (*Tylenchorynchus latus* Allen). Lance nematodes may feed either as ecto- or endoparasites, while the other migratory species are ectoparasites. Other species such as the root-lesion nematode *Pratylenchus brachyurus* (Godfrey) Filip. & Schuur.-Stekh., a migratory endoparasite, also feed extensively on cotton roots, but cause no apparent loss of yield or quality. All of the migratory species, as well as *M. incognita*, are favored by coarse soils. Lance and sting nematodes are primarily problems in the acidic sandy soils of the

Southeastern USA and the Mississippi River Valley, where they often occur in association with Fusarium wilt. Stunt nematode is an important migratory nematode pest of cotton in sandy soils of the United Arab Republic.

The interactions of nematodes and nutrient elements are particularly important, because nematodes usually occur in sandy soils which are naturally deficient in several nutrients; and because the nematode, by feeding on roots, may further aggravate nutrient deficiencies, predisposing the plant to other diseases and pests. In general, nutrient deficiencies of plants increase the severity of nematode diseases but decrease nematode populations per volume of soil. Correcting deficiencies often not only increases the plant's tolerance for the nematode, but also increases the reproduction of the nematode per gram of soil. This is due largely to the increased volume of the root system, but also might involve improved nutrition of the nematode. Specific examples that illustrate these generalities will be discussed.

On a sandy loam soil, N, P, and K fertilizers (alone or combined) increased cotton yields and reproduction of the stunt nematode, *T. latus*, and of *Pratylenchus* spp. (110,112). Complete fertilizer (NPK) at double the normal rate gave the greatest yield and tolerance to nematode damage; however, it caused a 51% increase in the number of *T. latus* per volume of soil. Among the single-element fertilizers, K gave the greatest increases both in tolerance to the nematode, and in reproduction of the nematode per volume of soil. When plants were deficient for K, nematode damage was greatly increased despite the low rate of reproduction of the nematodes.

A mixture of N-P-K (3-7-6) and microelements (Fe, 1.0%; Zn, 0.5%; Mn, 0.5%) also significantly increased cotton yields about 25% and root galling by *M. incognita* about 45% in sandy soils in California (62). The microelements alone gave only about a 10% yield increase, but an equal increase in galling. Fertilization with K alone also has a remarkable beneficial effect on cotton growth in K-deficient soils infested with *M. incognita* (111). Reproduction rates of the nematode also are increased per volume of soil, but this effect apparently is due to the greatly increased amount of root growth, because the mean number of egg masses per gram of root was unchanged for four different rates of K fertilization.

In cotton the reniform nematode, *R. reniformis*, apparently is affected by K differently than other nematodes. As with other nematode diseases, K fertilization greatly improves cotton yields. However, the mean number of nematode egg masses per gram of root decreases as the K level is increased. Thus, K deficiency appears to favor reproduction of this nematode, and correcting the deficiency may actually increase host resistance, as well as tolerance, to the nematode. In this respect K has an effect against the nematode similar to that against fungal pathogens causing seedling and wilt diseases. These results also suggest that K fertilization may have different effects on nematode reproduction on different cultivars, depending on the genetic level of resistance to the nematode in the cultivar.

Numerous studies have been concerned with the effects of vesicular-arbuscular mycorrhizal fungi on nematode damage to cotton. These symbiotic fungi facilitate the uptake of P and probably minor elements by the plant, and are ubiquitous in cotton fields. Their population densities, and the specific species involved, vary from one location to another. Mycorrhizal species that increase the tolerance of cotton to *M. incognita* in severely or slightly P-deficient soils include *Gigaspora margarita* Becker & Hall (=*Endogone calospora* Nicol. & Gerd.) (125,126,151), *Glomus etunicatus* Becker & Gerd. (127), *Glomus mosseae* (Nicol. & Gerd.) Gerd. & Trappe (54), *Glomus intraradices* Smith & Schenck (151), and *Glomus faciculatum* (136). *Glomus mosseae* also increases tolerance to *R. reniformis* (148). In severely deficient soils fertilization with P also substantially increases the tolerance of the plant to the nematode, and both P fertilization and inoculation with the mycorrhizal fungi increase the reproduction of the nematode per gram of soil. In soils with moderate amounts of P, the mycorrhizal fungi generally increase tolerance to nematodes more than additional P does. In one case, adding more P to already moderate amounts increased the damage to the host, whereas the mycorrhizae increased tolerance to the disease (150). In this case, cotton leaf tissue from the control or nematode-infested plots showed a deficient level of Zn (7 mg/kg), whereas leaves from the mycorrhizal plots contained 70% more Zn (12 mg/kg). Thus, the added P and nematodes may have aggravated a Zn deficiency which was corrected by the mycorrhizal fungi.

Some mycorrhizae may be antagonistic to nematodes apart from their effects on plant nutrition. *Glomus* species consistently reduce the reproduction of *M. incognita* per gram of cotton root (54,127,136,150), and *G. mosseae* reduces *R. reniformis* penetration by 27-60% and adult females by 63-87% (148) *Gigaspora margarita* similarly reduces the reproduction of *P. brachyurus* per gram of cotton root (53). Thus, mycorrhizae may be even more valuable than fertilization for increasing cotton resistance to certain nematodes. However, the effects of mycorrhizae on microelements must be defined better before this conclusion can be reached with certainty.

Although nematodes are widely known to predispose cotton plants to various fungal diseases, very few efforts have been made to determine whether this is accomplished by affecting the nutrient status of the plant. Inoculation of four-day-old cotton seedlings with *M. incognita* caused a 10-20% decrease in the Ca and Mg contents of cotyledons from 10-day-old plants, but did not appreciably affect concentrations in the hypocotyl or root (22). Thus, it was concluded that changes in Ca or Mg nutrition are not involved in predisposition to the soreshin pathogen, *R. solani*. Only minimal Ca nutrition was supplied in these studies, and plants were not examined after several weeks. Also, other elements were not considered. More detailed studies are needed to determine the effects of nematodes on Ca and Mg concentrations in cotton tissues.

Both *M. incognita* and *R. reniformis* reduce the K content of cotton roots and leaves (111). The degree of the reduction, as percentage of the control, is generally inversely proportional to the amount of K in the growth medium, and is greater in leaves than in roots. *M. incognita* also causes a significantly greater reduction of K in the leaves than does *R. reniformis*. These observations indicate that reduction of K in cotton tissue may be one way that these nematodes predispose cotton to seedling diseases and wilts. This conclusion agrees with the observation that lower concentrations of K are required to reduce Fusarium wilt in genetically-resistant cultivars than in susceptible cultivars (33). The possibility that nematodes predispose plants to disease by aggravating or causing nutrient deficiencies warrants more attention in future studies.

ELEMENTS AND INTEGRATED PEST MANAGEMENT

Macroelement deficiencies have had a major impact on cotton diseases. In the 1930's, Fusarium wilt and nematodes were considered the major disease problems of cotton in the USA, and often were associated with "cotton rust" (K deficiency). The discovery that Fusarium wilt and nematode damage were aggravated by K deficiency and acid soils (Ca deficiency) led to the widespread use of lime and K fertilizers on lands infested with these pathogens. These practices, along with resistant varieties and soil fumigation for nematodes, reduced Fusarium wilt to a minor disease in the 1980's, and the disease was essentially nonexistent where these integrated pest-management methods were used.

Excess use of imbalanced fertilizers, along with excess irrigation water, have contributed to the aggravation of cotton disease. The use of N and P fertilizers and irrigation water on cotton in the USA increased rapidly beginning about 1945, and peaked around 1965-1970 (37). Disease incidence also increased during this period and peaked at the same time (Table 1). From 1960 to 1970 the use of high rates of N and P, without proper attention to K, Ca, and minor elements, undoubtedly contributed to the peak lossed from seedling disease, Verticillium wilt, and boll rots. The reduced use of N and P fertilizers and irrigation water on cotton, since the advent of high energy costs in the early 1970's, also had contributed to marked reductions in these same diseases, along with Fusarium wilt and bacterial blight. While N rates have again increased somewhat in the 1980's, there had been a shift to use of N forms such as urea, that have fewer and smaller enhancing effects on diseases. Also, split applications of N have been refined to avoid effects of excess N, and much more attention is being given to maintaining required concentrations of other elements.

In general, any elemental deficiency enhances one or more diseases of cotton, and fertilizing to eliminate the deficiency decreases disease severity. However, fertilizers also may aggravate diseases, especially if they do not correct all of the nutrient deficiencies. For example, heavy use of N aggravates P, K, and Mg deficiencies, and in certain soils, liming may cause Zn deficiency. In turn, Zn deficiency, if left uncorrected, may cause P fertilizers to be toxic. Thus, all deficiencies of marginal levels of nutrients need to be identified by analysis of plants and soils. Also, analysis of plants should by used routinely to determine the effectiveness of any given fertilizer program. Many university soils laboratories now provide such analyses, and they should become even more common and economically feasible with advanced technology such as ion chromatography.

The eventual effectiveness of elemental fertilizers depends on the other integrated pest-management practices with which they are used. This principle was clearly illustrated in the early attempts to control Fusarium wilt with fertilizers in the Southeast. Such practices were not effective because of high concentrations of the fungus in the soil, use of highly susceptible cultivars, other elemental deficiencies, and the presence of large nematode populations. For optimum effectiveness, complete fertilizers should be used with resistant or tolerant cultivars and with minimal amounts of irrigation water for desired yields. In some cases, soil fumigation, chemical sprays, or crop rotation may be necessary to reduce the initial pest populations to a level that is responsive to fertilizers. When synergists such as nematodes are present, these may have to be controlled before the pathogen will show favorable responses to fertilizers. Finally excess N and P needs to be N-fixing bacteria, mycorrhizae, and antagonists of pathogens.

Future research should concentrate on several areas. First, the role of NH_3 in the toxicity of N fertilizers to the cotton plant and to soilborne pathogens should be better defined. It is obvious that NH_3 production in the soil may be beneficial or detrimental, depending on where, when, and in what concentration NH_3 occurs. More emphasis needs to be placed on taking advantage of the toxic properties of NH_3 to control pathogens. To do this, improved tolerance to NH_3 in cultivars should be developed. Also, beneficial microorganisms may be used to assure that NH_3 is formed at the right time and in the right place. Second, the effects of infection on the elemental status of plant tissues should be defined for more pathogens. The extensive K deficiency-disease complex in California apparently involves a pathogen in the development of K deficiency. Recent studies indicate that nematodes and *V. dahliae* may markedly lower K concentrations in tissues. Other root-nibbling pathogens such as *Pythium, Macrophomina* and *Thielaviopsis* also can be expected to have similar effects, but have not been studied. Third, the effects of Mg and minor elements (especially Zn and B) on disease severity need to be defined better, and Mg, Zn, and B amendments should be evaluated for disease control. Fourth, better methods should be developed to detect and treat elemental deficiencies in subsoils, and effects of subsoil deficiencies on diseases should be determined. Finally, better analytical techniques and data bases are needed to determine deficiencies based on petiole or foliar analyses. When this has been accomplished, future studies can determine the concentrations of elemental nutrients and elemental toxins in both the soil and plant tissue to gain a better understanding of the roles of elements in disease development. Such data would be invaluable for the optimum use of elemental fertilizers.

LITERATURE CITED

1. Abdel-Raheem, A., and Bird, L.S. 1967. Effect of nutrition on resistance and susceptibility of cotton to *Verticillium albo-atrum* and *Fusarium oxysporum* f. *vasinfectum*. Phytopathology 57:457.
2. Adams, F. 1966. Calcium deficiency as a causal agent of ammonium phosphate injury to cotton seedlings. Soil Sci. Soc. Am. Proc. 30:485-488.
3. Adams, J.E., Wilson, R.C., Hessler, L.E., and Ergle, D.R. 1939. Chemistry and growth of cotton in relation to soil fertility and root-rot. Proc. Soil Sci. Soc. Am. 4:329-332.
4. Albert, W.B. 1946. The effects of certain nutrient treatments upon the resistance of cotton to *Fusarium vasinfectum*. Phytopathology 36:703-716.
5. Ashworth, L.J., Jr., George, A.G., and McCutcheon, O.D. 1982. Disease-induced potassium deficiency and Verticillium wilt in

cotton. Calif. Agric. 36:18-20.

6. Ashworth, L.J., Jr., George, A.G., and M^CCutcheon, O.D. 1982. Disease-induced potassium deficiency and Verticillium wilt in cotton.Calif. Agric. 36:18-20.

7. Avizohar-Hershenzon, Z., and Shacked, P. 1969. Studies on the mode of action of inorganic nitrogenous amendments on *Sclerotium rolfsii* in soil. Phytopathology 59:288-292.

8. Baard, S.W., and Pauer, G.D.C. 1981. Effect of alternate drying and wetting of the soil, fertilizer amendment, and pH on the survival of microsclerotia of *Verticillium dahliae*. Phytophylactica 13:165-168.

9. Babaev, F.A. 1969. Some control measures against Verticillium wilt. Khlopkovodstvo 19:31-33.

10. Babaev, F.A., and Bagirov, M. 1967. On the control of Verticillium wilt of cotton. Zashch. Rast. (Moscow) 12:21-22.

11. Batson, W.E. 1971. Interrelationships among resistance to five major diseases and seed, seedling and plant characters in cotton. Ph.D. thesis, Texas A&M University, College Station. 126 pp.

12. Bell, A.A. 1967. Formation of gossypol in infected or chemically irritated tissues of *Gossypium* species. Phytopathology 57:759-763.

13. Bell, A.A. 1973. Nature of disease resistance. Pages 47-62 in: Verticillium Wilt of Cotton. U. S. Dept. Agric. Publ. ARS-S-19.

14. Bell, A.A. 1983. Protection practices in the USA and world. Section B- Diseases. Pages 288-309 in: Cotton. Agron. Monograph Series No. 24. R.J. Kohel and C.F. Lewis, eds. Am. Soc. Agron., Crop Sci. Soc. Am., and Soil Sci. Soc. Am., Madison, WI.

15. Bell, A.A., and Mace, M.E. 1979. Biochemistry and physiology of resistance. Pages 431-486 in: Fungal Wilt Diseases of Plants. M.E. Mace, A.A. Bell, and C.H. Beckman, eds. Academic Press, New York. J., Jr., and Huisman, O.C. 1976. Copper nutrition and development of Verticillium wilt disease. Proc. Am. Phytopathol. Soc. 3:314-315.

16. Blair, W.C., and Curl, E.A. 1974. Influence of fertilization regimes on the rhizosphere microflora and Rhizoctonia disease of cotton seedlings. Proc. Am. Phytopathol. Soc. 1:28-29.

17. Blank, L.M. 1941. Response of *phymatotrichum omnivorum* to certain trace elements. J. Agric. Res. 62:129-159.

18. Blank, L.M. 1944. Effect of nitrogen and phosphorus on the yield and root rot responses of early and late varieties of cotton. J. Am. Soc. Agron. 36:875-888.

19. Blank, L.M. 1962. Fusarium wilt of cotton moves west. Plant Dis. Rep. 46:396.

20. Blank, L.M., and Talley, P.J. 1941. Are ammonium salts toxic to the cotton root rot fungus? Phytopathology 31:926-935.
21. Bollenbacher, K., and Fulton, N.D. 1971. Effects of nitrogen compounds on resistance of *Gossypium arboreum* seedlings to *Colletotrichum gossypii*. Phytopathology 61:1394-1395.
22. Carter, W.W., Halloin, J.N., Hunter, R.E., Veech, J.A., and Crookshank, H.R. 1975. Distribution of calcium and magnesium in cotton seedlings infected by *Neloidogyne incognita* and *Rhizoctonia solani*. Pages 21-23 in: Proc. Beltwide Cotton Prod. Res. Conf., National Cotton Council, Memphis, TN.
23. Chan, B.G., and Wilhelm, S. 1984. A chemical investigation of a die-back cotton disease. Pages 29-31 in: Proc. Beltwide Cotton Prod. Res. Conf. J.M. Brown, ed. National Cotton Council, Memphis, TN.
24. Charudattan, R. 1970. Studies on strains of *Fusarium vasinfectum* Atk. Phytopathol. Z. 67:129-143.
25. Chernyayeva, I. I., Hasanova, F. S., and Muromtsev, G.S. 1984. Effect of various nitrogen forms upon *Verticillium dahliae* and the metabolism of some soil bacteria. Pages 703-710 in: Soil Biology and Conservation of Biosphere. J. Szegi, ed. Budape; Akademiai Kiado.
26. Chinnova, G.A. 1968. The spread of microorganisms antagonistic to *F. oxysporum* f. sp. *vasinfectum* in the root zone of cotton in relation to introduced fertilizers. Ispol'z. Mikroorgan. Zhivornov, Dlya Zasch. Rast. 50:144-149.
27. Christiansen, M.N., and Rowland, R.A. 1986. Germination and stand establishment. Pages 535-541 in: Cotton Physiology. J.R. Mauney and J. M. Stewart, eds. The Cotton Foundation, Memphis, TN.
28. Christensen, P.D., Stith, L.S., and Lyerly, P.J. 1954. The occurrence of Verticillium wilt in cotton as influenced by the level of salt in the soil. Plant. Dis. Rep. 38:309-310.
29. Clark, F.E. 1942. Experiments toward the control of the take-all disease of wheat and the Phymatotrichum root rot of cotton. US Dept. Agric. Tech. Bull. No. 835. 27pp.
30. Davis, R.M., Menge, J.A., and Erwin, D.C. 1979. Influence of *Glomus fasciculatus* and soil phosphorus on Verticillium wilt of cotton. Phytopathology 69:453-456.
31. Desai, D.B., and Wiles, A.B. 1976. Reaction foliar applications of microelements. Page 21 in: Proc. Beltwide Cotton Prod. Res. Conf. National Cotton Council, Memphis, TN.
32. Ebbels, D.L. 1975. Fusarium wilt of cotton: A review with special reference to Tanzania. Cotton Grow. Rev. 52:295-339.

33. El-Gindi, A.Y., Oteifa, B.A., and Khadr, A.S. 1974. Interrelationships of *Rotylenchulus reniformis, Fusarium oxysporum* f. *vasinfectum* and potassium nutrition of cotton, *Gossypium barbadense.* Potash Rev., Subj. 23:1-5.

34. El Nur, E., and Abdel Fattah, M.A. 1970. Fusarium wilt of cotton. Tech. Bull. No. 2 (New Series). Agri. Res. Corp. Bot. and Plant Pathol. Sec., Gezira Research Station, Wad Medani, Sudan. 16pp.

35. El-Zik, K.M. 1984. Integrated control of Verticillium wilt of cotton. Pages 53-54 in: Proc. Beltwide Cotton Prod. Res. Conf. J.M. Brown, ed. National Cotton Council, Memphis, TN.

36. El-Zik, K.M. 1985. Integrated control of Verticillium wilt of cotton. Plant Dis. 69:1025-1032.

37. El-Zik, K.M. 1986. Half a century dynamics and control of cotton disease: Dynamics of cotton diseases and their control. Pages 29-33 in: Proc. Beltwide Cotton Prod. Res. Conf. J.M. Brown and T.C. Nelson, eds. National Cotton Council, Memphis, TN.

38. Ezekiel, W.N. 1930. Report of the cotton-root-rot conference at Temple, Texas. Phytopathology 20:889-894.

39. Ezekiel, W.N., Taubenhaus, J.J., and Carlyle, E.C. 1930. Soil-reaction effects on Phymatotrichum root rot. Phytopathology 20:803-815.

40. Ezekiel, W.N., Taubenhaus, J.J., and Fudge, J.F. 1934. Nutritional requirements of the root-rot fungus *Phymatotrichum omnivorum.* Plant Physiol. 9:187-216.

41. Foy, C.D., Webb, H.W., and Jones, J.E. 1981. Adaptation of cotton genotypes to an acid, manganese toxic soil. Agron. J. 73:107-111.

42. Fraps, G.S., and Fudge, J.F. 1935. Relation of the occurrence of cotton root rot to the chemical composition of soils. Tes. Agric. Exp. Stn. Bull. No. 522. 21pp.

43. Fulton, H.R. 1927. Organic fertilizers and cotton wilt control. Science 66:193-194.

44. Grishechkina, L.D., Sidorova, S.F. 1984. Effect of mineral compounds on leaf infection by cotton wilt pathogen. Mikol Fitopatol. 18:66-70.

45. Hafez, A.A.R., Stout, P.R., and DeVay, J.E. 1975. Potassium uptake by cotton in relation to Verticillium wilt. Agron. J. 67:359-361.

46. Hinkle, D.A., and Brown, A.L. 1968. Secondary nutrients and micronutrients. Pages 281-320 in: Advances in Production and Utilization of Quality Cotton: Principles and Practices. F.C. Elliot, M. Hoover, and W.K. Porter, Jr., eds. Iowa State Univ. Press, Ames.

47. Hinkle, D.A., and Staten, G. 1941. Fertilizer experiments with Acala cotton on irrigated soils. N.M. Agric. Exp. Stn. Bull. 280. 15pp.

48. Howard, D.D., and Adams, F. 1965. Calcium requirement for penetration of subsoils by primary cotton roots. Soil Sci. Soc. Am. Proc. 29:558-562.

49. Howell, C.R. 1976. Use of enzyme-deficient mutants of *Verticillium dahliae* to assess the importance of pectolytic enzymes in symptom expression of Verticillium wilt of cotton. Physiol. Plant Pathol. 9:279-283.

50. Howell, C.R., Beier, R.C., and Stipanovic, R.D. 1988. Production of ammonia by *Enterobacter clocae* and its possible role in the biological control of Pythium preemergence damping-off by the bacterium. Phytopathology 78:1075-1078.

51. Howell, C.R., and Stipanovic, R.D. 1987. Production of a volatile antibiotic by *Enterobacter cloacae* and its possible role in the biological control of pathogenic fungi by the bacterium. Phytopathology 75:1720.

52. Huang, B.F., and M^{C}Master, B.J. 1985. Ammonia evolution as an antifungal mechanism *in vitro*. Phytopathology 75:1360

53. Hussey, R.S., and Roncadori, R.W. 1978. Interaction of *Pratylenchus brachyurus* and *Gigaspora margarita* on cotton. J. Nematol. 10:16-20.

54. Hussey, R.S., and Roncadori, R.W., 1982. Vesicular-arbuscular mycorrhizae may limit nematode activity and improve plant growth. Plant Dis. 66:9-14.

55. Isaev, B.M. 1963. Microelements and wilt. Khlopkovodstvo 13:44-45.

56. Isaev, B.M. 1966. Importance microelements in the control of Verticillium wilt of cotton. Tr. Vses. Nauchno-Issled. Inst. Khlopkovod. 9:86-90.

57. Jodidi, S.L. 1912. Amino acids and acid amides as source of ammonia in soils. Pages 327-362 in: Iowa State Agric. Exp. Stn. Bull. No. 9.

58. Joham, H.E. 1957. Carbohydrate distribution as affected by calcium deficiency in cotton. Plant Physiol. 32:113-117.

59. Joham, H.E. 1971. Some aspects of the zinc nutrition of cotton. Pages 44-45 in: Proc. Beltwide Cotton Prod. Res. Conf. National Cotton Council, Memphis, TN.

60. Jordan, H.V., Dawson, P.R., Skinner, J.J., and Hunter, J.H. 1934. The relation of fertilizers to the control of cotton root rot in Texas. U.S. Dept. Agric. Tech. Bull. No. 426. 76pp.

61. Jordan, H.V., Nelson H.A., and Adams, J.E. 1939. Relation of

fertilizers, crop residues, and tillage to yields of cotton and incidence of root rot. Proc. Soil Sci. Soc. Am. 4:325-328.

62. Jorgenson, E.C. 1984. Nematicides and nonconventional soil amendments in the management of root-knot nematode on cotton. J. Nematol. 16:154-158.

63. Ju-Shen, H., and Tsing-Hai, C. 1957. The protective effect of some growth-stimulating substances and minor elements on cotton seedling diseases. Acta Phytopathol. Sin. 3:190-191.

64. Kalyanasundaram, R., and Saraswathi-Devi, L. 1955. Zinc in the metabolism of *Fusarium vasinfectum* Atk. Nature 176:945.

65. Kaufman, D. D., and Williams, L.E. 1965. Influence of soil reaction and mineral fertilization on numbers and types of fungi antagonistic to four soil-borne plant pathogens. Phytopathology 55:570-574.

66. Kent, L.M., and Lauchli, A. 1985. Germination and seedling growth of cotton *gossypium hirsutum*:: Salinity-calcium interactions. Plant Cell Environ. 8:135-160.

67. Kesavan, R., Prasad, N.N. 1975. Effect of certain carbon and nitrogen sources *in vitro* production of fusaric acid by muskmelon wilt pathogen. Indian Phytopathol. 28:29-32.

68. King, C.J., Beckett, R.E., and Parker, O. 1938. Agricultural investigations at the United States Field Station, Sacaton, Ariz., 1931-1935. U S Dept. Agric. Circ. No. 479. 64pp

69. King, C.J., Hope, C., and Eaton, E.D. 1934. Some microbial activities affected in menurial control of cotton root rot. J. Agric. Res. 49:10931107.

70. Lehle, F.R., and Hofmann, W.C. 1987. Calcium stimulation of radicle growth of cotton seed imbibed at a suboptimal temperature. Page 85 in: Proc. Beltwide Cotton Prod. Res. Conf. J. M. Brown and T. C. Nelson, eds. National Cotton Council, Memphis, TN.

71. Lewis, A.C. 1911. Wilt disease of cotton in Georgia and its control. Ga. State Board Entomol. Bull. No. 34. 29 pp.

72. Logan, C. 1958. Some observations and experiments on Abyan root-rot of cotton. Emp. Cotton Grow. Rev. 35:168-175.

73. Longenecker, D.E., and Hefner, J.J. 1961. Effect of fertility level, soil moisture and trace elements on the incidence of Verticillium wilt in Upland irrigated cotton. Tex. Agric. Exp. Stn. Prog. Rep. 2175. 5 pp.

74. Lyda, S.D. 1971. Adjusting sodium content of soils to control Phymatotrichum root rot. Pages 555-564 in: International Symposium Use of Isotopes and Radiation in Agriculture and Animal Husbandry Research. Indian Agric. Res. Inst.,

New Delhi.

75. Lyda, S.D. 1973. Studies on *Phymatotrichum omnivorum* and Phymatotrichum root rot. Pages 69-73 in: The Relation of Soil Microorganisms to Soilborne Plant Pathogens. South. Coop. Ser. Bull. 183.

76. Lyda, S.D. 1975. Environmental factors influencing the distribution and survival of *Phymatotrichum omnivorum.* Pages 165-168 in: Proc. Beltwide Cotton Prod. Res. Conf. National Cotton Council, Memphis, TN.

77. Lyda, S.D. 1978. Ecology of *Phymatotrichum omnivorum.* Annu. Rev. Phytopathol. 16:193-209.

78. Lyda, S.D., and Kissel, D.E. 1974. Sodium influence on disease development and sclerotial formation by *Phymatotrichum omnivorum.* Proc. Am. Phytopathol. Soc. 1:163-164.

79. McLean, H.C., and Wilson, G.W. 1914. Ammonification studies with soil fungi. N.J. Agric. Exp. Stn. Bull. No. 270. 38 pp.

80. Mace, M.E., Stipanovic, R.D., and Bell, A.A. 1985. Toxicity and role of terpenoid phytoalexins in Verticillium wilt resistance in cotton. Physiol. Plant Pathol. 26:209-218.

81. Malca, I., Erwin, D.C., Moje, W., and Jones, B. 1966. Effect of pH and carbon and nitrogen sources on the growth of *Verticillium albo-atrum.* Phytopathology 56:401-406.

82. Marcus-Wyner, L., and Rains, D.W. 1982. Nutritional disorders of cotton plants. Commun. Soil Sci. Plant Anal. 13:685-736.

83. Matocha, J.E., and Bird, L.S. 1984. Effects of certain cultural and chemical treatments on cotton root rot control. Pages 35-36 in: Proc. Beltwide Cotton Prod. Res. Conf. J.M. Brown, ed. National Cotton Council, Memphis, TN.

84. Matocha, J.E., and Heilman, M.D. 1982. Cotton response to soil treatments for root rot control. Pages 32-33 in: Proc. Beltwide Cotton Prod. Res. Conf. J.M. Brown, ed. National Cotton Council, Memphis, TN.

85. Matocha, J.E., Mostaghimi, S., and Hopper, F.L. 1986. Effect of soil and plant treatments on Phymatotrichum root rot. Pages 24-26 in: Proc. Beltwide Cotton Prod. Res. Conf. J.M. Brown and T.C. Nelson, eds. National Cotton Council, Memphis, TN.

86. Mekhmanov, M. 1967. Effect of zinc and molybdenum on cotton resistance to Verticillium wilt. Uzb. Biol. Zh. 11:47-50.

87. Mekhmanov, M., and Pirakjunov, T. 1968. The effect of nitrogenous fertilization on cotton resistance to *Verticillium dahliae.* Uzb. Biol. Zh. 3:29-32.

88. Miles, L.E. 1936. Effect of potash fertilizers on cotton wilt. Miss. Agric. Exp. Stn. Tech. Bull. No. 23. 21pp.

89. Miller, R.W., Nigh, E.L., Jr., and Alcorn, S.M. 1967. Nematodes associated with *Verticillium albo-atrum* attacking irrigated cotton in the Southwest. Page 52 In: Proc. Beltwide Cotton Prod. Res. Conf. National Cotton Council, Memphis, TN.

90. Mirpulatova, N.S. 1961. Some results from research on wilt. Khlopkovodstvo 11:30-33.

91. Mostaghimi, S., Matocha, J.E., Branick, P.S., El-Zik, K., and Bird, L. S. 1987. Suppression of Phymatotrichum root rot through chemical and biological treatments of soil and plant. Pages 35-37 in: Proc. Beltwide Cotton Prod. Res. Conf. J.M. Brown and T.C. Nelson, eds. National Cotton Council, Memphis, TN.

92. Mozumder, B.K.G., and Caroselli, N.E. 1974. Influence of microelements in a sporulation medium on subsequent water requirements for germination of *Verticillium albo-atrum* conidia. Indian Phytopathol. 27:537-541.

93. Mueller, J.P., Hine, R.B., Pennington, D.A., and Ingle, S.J. 1983. Relationship of soil cations to the distribution of *Phymatotrichum omnivorum* Phytopathology 73:1365-1368.

94. Mussell, H.W. 1975. Endopolygalacturonase: Evidence for involvement in Verticillium wilt of cotton. Phytopathology 63:62-70.

95. Mustakimova, F., and Muratmukhamedov, K.H. 1982. Increasing cotton wilt resistance due to the effect of zinc sulfate and hairy vetch. Izv. Akad. Nauk Turkm. SSR, Ser. Biol. Nauk 0:67-70.

96. Naim, M.S., and Shaaban, A.S. 1965. Relation of nitrogen fertilizers to growth vigour of *Fusarium* wilted cotton. Phytopathol. Mediterr. 4:145-153.

97. Naim, M.S., and Sharoubeem, H.H. 1964. Carbon and nitrogen requirements of *Fusarium oxysporum* causing cotton wilt. Mycopathologia 22:59-64.

98. Naim, M.S., Sharoubeem, H.H., and Habib, A.A. 1966. The relation of different nitrogen levels to the incidence of vascular wilt and growth vigour of Egyptian cotton. Phytopathol. Z. 55:257-264.

99. Nakamura, K., Nishima, A.M., Fereira, M.E., Banzatto, D.A., and Kronka, S.N. 1975. Efeito de diferentes niveis de N, P e K a adicionados ao solo sobre a murcha de Fusarium do algodoeiro (*Gossypium hirsutum* L.) causada por *Fusarium oxysporum* f. *vasinfectum* (Atk) Snyder & Hansen. Cientifica 3:174-186.

100. Neal, D.C. 1927. Cotton Wilt: A pathological and physiological investigation. Ann. Mo. Bot. Gard. 14:359-407.

101. Neal, D.C. 1928. Cotton diseases in Mississippi and their control. Miss. Agric. Exp. Stn. Bull. No. 248. 30pp.

102. Neal, D.C. 1935. Further studies on the effect of ammonia nitrogen on growth of the cotton-root-rot fungus, *Phymatotrichum omnivorum*, in field and laboratory experiments. Phytopathology 25:967-968.
103. Neal, D.C., and Collins, E.R. 1936. Concentration of ammonia necessary in a low-lime phase of Houston clay soil to kill the cotton root-rot fungus. *Phymatotrichum omnivorum*. Phytopathology 26:1030-1032.
104. Neal, D.C., and Sinclair, J.B. 1960. Nitrogen supplement in the Louisiana-Mississippi River Delta as a possible control for Verticillium wilt of cotton. Plant Dis. Rep. 44:478.
105. Neal, D.C., Wester, R.E., and Gunn, K.C. 1932. Treatment of cotton root-rot with ammonia. Science 75:139-140.
106. Neal, D.C., Wester, R.E., and Gunn, K.C. 1933. Growth of the cotton root rot fungus in synthetic media, and the toxic effects of ammonia on the fungus. J. Agric. Res. 47:107-118.
107. Office of Governmental and Public Affairs. 1987. 1987 Fact Book of U S. Agriculture. U S Dept. Agric. Misc. Publ. 1063. 163pp.
108. Orellana, R.G., Foy, C.D., and Fleming, A.L. 1975. Effect of soluble aluminum on growth and pathogenicity of *Verticillium albo-atrum* and *Whetzelinia sclerotiorum* from sunflower. Phytopathology.
109. Orton, W.A. 1908. Cotton wilt. U S Dept. Agric. Farmer's Bull. 333. 24pp
110. Oteifa, B.A., and Diab, K.A. 1961. Significance of potassium fertilization in nematode infested cotton fields. Plant Dis. Rep. 45:932.
111. Oteifa, B.A., and Elgindi, A.Y. 1976. Potassium nutrition of cotton, *Gossypium barbadense*, in relation to nematode infection by *Meloidogyne incognita* and *Rotylenchus reniformis*. Pages 301-306 in: Fertilizer Use and Plant Health. International Potash Institute, Bern, Switzerland.
112. Oteifa, B.A., Elgindi, D.M., and Diab, K.A. 1965. Cotton yield and population dynamics of the stunt nematode, *Tylenchorhynchus latus* under mineral fertilization trials. Potash Rev., Subj. 23:1-7.
113. Pinckard, J.A., and Leonard, O.A. 1944. Influence of certain soil amendments on the yield of cotton affected by the *Fusarium heterodera* complex. J. Am. Soc. Agron. 36:829-843.
114. Presley, J.T. 1950. Verticillium wilt of cotton with particular emphasis on variation of the causal organism. Phytopathology 40:497-511.
115. Presley, J.T., and Dick, J.B. 1951. Fertilizer and weather affect

Verticillium wilt. Miss. Farm Res. 14:1-6.

116. Presley, J.T., and Leonard, O.A. 1948. The effect of calcium and other ions on the early development of the radicle of cotton seedlings. Plant Physiol. 23:516-525.

117. Puente, F. 1965. Some effects of soil temperature and phosphorus and calcium levels on cotton seedling growth. Diss. Abstr. 26:4157.

118. Puhalla, J.E., and Hummel, M. 1983. Vegetative compatibility groups within *Verticillium* *dahliae*. Phytopathology 73:1305-1308.

119. Ramasami, R., and Shanmugam, N. 1976. Effect of nutrients on the incidence of Rhizoctonia seedling disease of cotton. Indian Phytopathol. 29:465-466.

120. Ramasamy, K., and Prasad, N.N. 1975. Potassium and melon wilt. III. Influence of potassium on *in vitro* fusaric acid and pathogen's propagule. Sci. Cult. 41:525-526.

121. Ranney, C.D. 1962. Effects of nitrogen source and rate on the development of Verticillium wilt of cotton. Phytopathology 52:38-41.

122. Ranney, C.D., and Bird, L.S. 1958. Influence of fungicides, calcium salts, growth regulators and antibiotics on cotton seedling disease when mixed with the covering soil. Plant Dis. Rep. 42:785-791.

123. Rast, L.E. 1924. Control of cotton wilt by the use of potash fertilizers. J. Am. Soc. Agron. 14:222-223.

124. Reynolds, E.B., and Rea, H.E. 1934. Effect of fertilizers on yields of cotton and on the control of the root-rot disease of cotton on the Blackland prairie soils of Texas. J. Am. Soc. Agron. 26:313-318.

125. Rich, J.R., and Bird, G.W. 1974. Association of early-season vesicular-arbuscular mycorrhizae with increased growth and development of cotton. Phytopathology 64:1421-1425.

126. Roncadori, R.W., and Hussey, R.S. 1977. Interaction of the endomycorrhizal fungus *Gigaspora margarita* and root-knot nematode on cotton. Phytopathology 67:1507-1511.

127. Roncadori, R.W., and Hussey, R.S. 1979. Interaction of *Glomus etunicatus*, a vesicular-arbuscular mycorrhizal fungus, and *Meloidogyne incognita* on cotton. Page 17 in: Proc. Beltwide Cotton Prod. Res. Conf. J.M. Brown, ed. National Cotton Council, Memphis, TN.

128. Roncadori, R.W., McCarter, S.M., and Crawford, J.L. 1975. Evaluation of various control measures for cotton boll rot. Phytopathology 65:567-570.

129. Rouse, R.D., and Adams, F. 1958. Lime for good stand and yield

of cotton. Highlights Agric. Res., Ala. Agric. Exp. Stn. 5:11.

130. Rudolph, B.A., and Harrison, G.J. 1939. Attempts to control Verticillium wilt of cotton and breeding for resistance. Phytopathology 29:752.
131. Rush, C.M., and Lyda, S.D. 1978. Anhydrous ammonia fumigation of *Phymatotrichum*-infested montmorillonitic clay. Pages 23-25 in: Proc. Beltwide Cotton Prod. Res. Conf. J. M. Brown, ed. National Cotton Council, Memphis, TN.
132. Rush, C. M., and Lyda, S. D. 1982. Effects of anhydrous ammonia on mycelium and sclerotia of *Phymatotrichum omnivorum*. Phytopathology 72:1085-1089.
133. Rush, C.M., and Lyda, S.D. 1984. Evaluation of deep-chiseled anhydrous ammonia as a control for Phymatotrichum root rot of cotton. Plant Dis. 68:291-293.
134. Sadasivan, T.S. 1965. Effect of mineral nutrients on soil microorganisms and plant disease. Pages 460-470 in: Ecology of Soil-borne Plant Pathogens: Prelude to Biological Control. K. F. Baker and W. C. Snyder, eds. Univ. California Press, Berkeley.
135. Sadasivan, T.S., and Saraswathi-Devi, L. 1957. Vivotoxins and uptake of ions by plants. Curr. Sci. (India) 26:74-75.
136. Saleh, H., and Sikora, R.A. 1984. Relationship between *Glomus fasciculatum* root colonization of cotton and its effect on *Meloidogyne incognita*. Nematologica 30:230-237.
137. Savov, S.G. 1986. Effectiveness of some trace elements in the control of cotton Verticillium wilt. Rastenievud. Nauki 23:68-71.
138. Schonbeck, F., and Dehne, H.W. 1977. Damage to mycorrhizal and nonmycorrhizal cotton seedlings by *Thielaviopsis basicola*. Plant Dis. Rep. 61:266-267.
139. Selvaraj, J.C. 1974. Growth response of the isolates of *Verticillium* to nitrogen nutrition. Indian Phytopathol. 27:434-437.
140. Selvaraj, J.C. 1974. Alteration of the production and activity of pectinases of *Verticillium dahliae* by calcium chloride and sodium chloride. Indian Phytopathol. 27: 437-441.
141. Shao, F.M., and Foy, C.D. 1982. Interaction of soil manganese and reaction of cotton to Verticillium wilt and Rhizoctonia root rot. Commun. Soil Sci. Plant Anal. 13:21-38.
142. Sharoubeem, H.H., Naim, M.S., and Habib, A.A. 1965. Phosphorus and *Fusarium* in sand cultures in relation to growth-vigour of certain Egyptian cotton varieties. Phytopathol. Mediterr. 4:85-95.
143. Sharoubeem, H.H., Naim, M.S., and Habib, A.A. 1966. Interaction of phosphorus and *Fusarium* with the major element

nutrition of cotton plants. Acta Phytopathol. Acad. Sci. Hung. 1:53-65.

144. Sharoubeem, H.H., Naim, M.S., and Habib, A.A. 1966. Effects of different levels of potassium, alone or in combination with *Fusaria* spp., on the nutritional status of cotton plants. Acta Phytopathol. Acad. Sci. Hung. 1:208-222.

145. Sharoubeem, H.H., Naim, M.S., and Habib, A.A. 1966. Effect of different levels of potassium on growth-vigour of cotton variety plants in relation to *Fusarium* spp. associated with the vascular wilt disease. Mycopthol. Mycol. Appl. 29:65-81.

146. Sharoubeem, H.H., Naim, M.S., and Habib, A.A. 1967. Combined effect of nitrogen supply and *Fusarium* infection on the chemical composition of cotton plants. Acta Phytopathol. Acad. Sci. Hung. 2:39-48.

147. Sharoubeem, H.H., Naim, M.S., and Habib, A.A. 1967. Potassium, nitrogen, and phosphorus in relation to the incidence of cotton wilt caused by *Fusarium oxysporum* f. *vasinfectum*. J. Microbiol. U.A.R. 2:1-8.

148. Sikora, R.A., and Sitaramaiah, K. 1980. Antagonistic interaction between the endotrophic mycorrhizal fungus *Glomus mosseae* and *Rotylenchulus reniformis* on cotton. Nematropica 10:72-73.

149. Smiley, R.W., Cook, R.J., and Papendick, R.I. 1970. Anhydrous ammonia as a soil fungicide against *Fusarium* and fungicidal activity in the ammonia retention zone. Phytopathology 60:1227-1232.

150. Smith, G.S., Roncadori, R.W., and Hussey, R.S. 1985. Development of *Meloidogyne incognita* on cotton as affected by the endomycorrhizal fungus, *Glomus intraradices*, and phosphorus. J. Nematol. 17:514.

151. Smith, G.S., Roncadori, R.W., and Hussey, R.S. 1986. Interaction of endomycorrhizal fungi, superphosphate, and *Meloidogyne incognita* on cotton in microplot and field studies. J. Nematol. 18:208-216.

152. Smith, R.B., and Hallmark, C.T. 1987. Selected chemical and physical properties of soils manifesting cotton root rot. Agron. J. 79:155-159.

153. Soileau, J.M., Engelstad, O.P., and Martin, J.B., Jr. 1969. Cotton growth in an acid fragipan subsoil: II. Effects of soluble calcium, magnesium and aluminum on roots and tops. Soil Sci. Soc. Am. Proc. 33:919-924.

154. Stapelton, J.J., DeVay, J.E., and Lear, B. 1987. Effect of combining ammonia-based fertilizers with solarization on pathogen control

and plant growth. Phytopathology 77:1744.

155. Streets, R.B. 1937. Phymatotrichum (Cotton or Texas) root rot in Arizona. Pages 298-410 in: Ariz. Agric. Exp. Stn. Tech. Bull. No. 71.

156. Streets, R.B., and Bloss, H.E. 1973. Phymatotrichum root rot. Monograph No. 8. The American Phytopathological Society, St. Paul, MN. 38 pp.

157. Talley, P.J., and Blank, L.M. 1941. A critical study of the nutritional requirements of *Phymatotrichum omnivorum*. Plant Physiol. 16:1-18.

158. Talley, P.J., and Blank, L.M. 1942. Some factors influencing the utilization of inorganic nitrogen by the root rot fungus. Plant Physiol. 17:52-68.

159. Taubenhaus, J.J., and Ezekiel, W.N. 1930. Recent studies on Phymatotrichum root-rot. Am. J. Bot. 17:554-571.

160. Taubenhaus, J.J., Ezekiel, W.N., and Fudge, J.F. 1937. Relation of soil acidity to cotton root rot. Tex. Agric. Exp. Stn. Bull. No. 545. 39 pp.

161. Taubenhaus, J.J., Ezekiel, W.N., and Killough, D.T. 1928. Relation of cotton root rot and Fusarium wilt to the acidity and alkalinity of the soil. Tex. Agric. Exp. Stn. Bull. No. 389. 19 pp.

162. Throneberry, G.O. 1973. Some physiological responses of *Verticillium albo-atrum* to zinc. Can. J. Bot. 51:57-59.

163. Tisdale, H.B., and Dick, J.B. 1939. The development of wilt in a wilt-resistant and in a wilt-susceptible variety of cotton as affected by N-P-K ratio in fertilizer. Proc. Soil. Sci. Soc. Am. 4:333-334.

164. Tisdale, H.B., and Dick, J.B. 1942. Cotton wilt in Alabama as affected by potash supplements and as related to varietal behavior and other important agronomic problems. J. Am. Soc. Agron. 34:405-425.

165. Veech, J.A. 1984. Cotton protection practices in the USA and world. Section C - Nematodes. Pages 309-330 in: Cotton. Agron. Monograph Ser. No. 24. R. J. Kohel and C. F. Lewis, eds. Amer. Soc. Agron., Crop Sci. Soc. Amer., and Soil Sci. Soc. Amer., Madison, WI.

166. Walker, M.N. 1930. Potash in relation to cotton wilt. Fla. Agric. Exp. Stn. Bull. No. 213. 10 pp.

167. Ware, J.O., and Young, V.H. 1934. Control of cotton wilt and "rust." Arkansas Agric. Exp. Stn. Bull. No. 308. 23 pp.

168. Weinhold, A.R., Bowman, T., and Dodman, R.L. 1969. Virulence of *Rhizoctonia solani* as affected by nutrition of the pathogen. Phytopathology 59:1601-1605.

169. Weinhold, A.R., Dodman, R.L., and Bowman, T. 1972. Influence of exogenous nutrition on virulence of *Rhizoctonia solani*. Phytopathology 62:278-281.

170. Weir, B., DeVay, J.E., Garber, R.H., Stapleton, J.J., Felix, R., and Wakeman, R. J. 1988. Verticillium wilt-bronzing leaf syndrome potassium deficiency complex in cotton. Calif. Agric. (in press).

171. Weir, B.L., DeVay, J., Stapleton, J., Wakeman, J., and Garber, D. 1987. Effect of solarization on potassium disease complex of cotton. Pages 51-53 in: Proc. Beltwide Cotton Prod. Res. Conf. J. M. Brown and T. C. Cotton, eds. National Cotton Council, Memphis, TN.

172. Wiles, A.B. 1959. Calcium deficiency in cotton seedlings. Plant Dis. Rep. 43:365-367.

173. Wilhelm, S. 1950. The inoculum potential of *Verticillium albo-atrum* as affected by soil amendments. Phytopathology 40:970.

174. Woltz, S.S., and Jones, J.P. 1968. Micronutrient effects on the growth and pathogenicity of *Fusarium oxysporum* f. sp. *lycopersici*. Phytopathology 58:336-338.

175. Young, P.A. 1943. Cottons resistant to wilt and root knot and the effect of potash fertilizer in East Texas. Tex. Agric. Exp. Stn. Bull. No. 627. 26 pp.

176. Young, V.H., Fulton, N.D., and Waddle, B.A. 1959. Factors affecting the incidence and severity of Verticillium wilt disease of cotton. Arkansas Exp. Stn. Bull. No. 612. 26 pp.

177. Young, V.H., Janssen, G., and Ware, J.O. 1932. Cotton wilt studies. IV. Effect of fertilizers on cotton wilt. Arkansas Agric. Exp. Stn. Bull. No. 272. 27 pp.

178. Young, V.H., and Tharp, W.H. 1941. Relation of fertilizer balance to potash hunger and the fusarium wilt of cotton. Arkansas Agric. Exp. Stn. Bull. No. 410. 24 pp.

179. Zyngas, J.P. 1962. The effect of plant nutrients and antagonistic microorganisms on the damping-off of cotton seedlings caused by *Rhizoctonia solani* Kuhn. Diss. Abstr. 23:3587.

EVIDENCE FOR THE ROLE OF CALCIUM IN REDUCING ROOT DISEASE INCITED BY *PYTHIUM* SPP

Wen-Hsiung Ko and Ching-Wen Kao
Beaumont Agricultural Research Center
University of Hawaii
Hilo, Hawaii 96720

Members of the genus *Pythium* are present in agricultural and forest lands all over the world. Many of them are soilborne pathogens capable of causing serious economic losses on a wide range of hosts (13,14). It has been long observed that soils from some areas were not favorable for the establishment or survival of certain *Pythium* pathogens (34). The lower incidence of a soilborne disease in a given field in comparison with others in the nearby area also has been observed frequently (5); however, there have been relatively few and sporadic reports concerning the control of *Pythium* diseases with Ca in crop production (Table 1). In 1940, Vanterpool (36) showed that gypsum ($CaSO_4$) inhibited the browning root rot of wheat caused by *Pythium arrhenomanes* Drechsler and *Pythium tardicrescens* Vanterpool. Since this first report, Angell (2) reported that liming steamed soil to which *Pythium ultimum* Trow was added reduced seedling blight of opium poppy and Garren (9,12) reported the application of gypsum decreased peanut pod rot caused by *Pythium myriotylum* Drechsler. Gill (11) reported the addition of gypsum to soils infested with *P. myriotylum* or *Pythium aphanidermatum* (Edson) Fitzp. increased the percent of healthy seedlings of seven kinds of ornamental and vegetable plants over the control and Lewis and Lumsden (28) showed that CaO significantly reduced damping-off of peas caused by *P. ultimum* in naturally-infested soils in both greenhouse and field studies. Calcium oxide also reduced damping-off of soybeans, peppers, sugarbeets and beans caused by the same pathogen in their greenhouse tests. Kao and Ko (17) found that damping-off of cucumber seedlings caused by *Pythium splendens* Braun in artificially infested soil was decreased about 50% by the application of lime in both greenhouse and field tests.

Table 1. Examples of successful application of calcium to the control of soil-borne diseases caused species of *Pythium*.

Disease	Host	Pathogen	Year	Ref
Browning root rot	Wheat (*Triticum aestimum* L.)	*P. arrhenomanes* Drechsler	1940	36
Browning root rot	Wheat	*P. tardicrescens* Vanterpool		
Seedling blight	Poppy (*Papaver somniferum* L.)	*P. ultimum* Trow	1950	2
Pot Rot	Peanut (*Arachis hypogea* L.)	*P. myriotylum* Drechsler	1964 1968	9 12
Damping-off	Pea (*Pisum sativum* L.)	*P. ultimum*	1984	28
Damping-off	Soybean (*Glycine max* L.)	*P. ultimum*	1984	28
Damping-off	Pepper (*Capsicum annum* L.)	*P. ultimum*	1984	28
Damping-off	Sugarbeet (*Beta vulgaris* L.)	*P. ultimum*	1984	28
Damping-off	Bean (*Phaseolus vulgaris* L.)	*P. ultimum*	1984	28
Damping-off	Cucumber (*Cucumis sativus* L.)	*P. splendens* Braun	1986	17
Seedling disease	Tomato (*Lycopersicon esculentum* Mill.)	*P. myriotylum* or *P. aphanidermatum*	1972	11
Seedling disease	Onion (*Allium cepa* L.)	(Edson)Fitzp.	1972	11
Seedling disease	Stock (*Matthiola incana* (L.) R. Br.		1972	11
Seedling disease	Pansy (*Viola tricolor* L.)		1972	11
Seedling disease	Snapdragon (*Antirrhinum majus* L.)		1972	11
Seedling disease	Sweet William (*Dianthus barbatus* L.)		1972	11
Seedling disease	Calendula (*Calendula officinalis* L.)		1972	11

There was no consistent effect of Ca on saprophytic growth of *P. ultimum* in California soils suppressive or conducive to Pythium disease (29).

PYTHIUM SUPPRESSIVE SOILS

Pathogen and disease-suppressive soils may be relatively common in nature and also may exist in localized areas in a conducive field (33). Fungistasis potential varies greatly among different soils (8) and existence of volatile or non-volatile inhibitory substances in certain soils also has been reported (14,23,24,25). A simple method was developed for assaying small amounts of soil for suppressiveness to *Pythium splendens* Braun, one of the most destructive pathogens of cucumber in Hawaii (21). Sporangia were placed on the smooth surface of a soil block (ca. 8 g) on a glass slide and spore germination was determined after incubation directly on the soil surface with a vertical illumination microscope (18). Sporangia of *P. splendens* were suspended in cucumber root extract before being added to the soil surface to overcome general soil fungistasis. Soils were collected from locations representing different vegetation, soil type, or elevation. Thirty-seven percent of the 81 soil samples collected from the islands of Hawaii, Maui, Molokai and Kauai were suppressive to *P. splendens* in that they supported less than 50% germination (26). There was an inverse relationship between the percentage of sporangial germination and soil pH. Near neutral to alkaline soils in Hawaii were more suppressive to *P. splendens* than acid soils. A pasture soil (very fine sandy loam, pH 6.8) from the South Kohala district on the island of Hawaii was selected for further study because it was highly suppressive to

Table 2. Germination and damping-off of cucumber seedlings, and population of *Pythium splendens* in soil.

| Soil | *P. Splendens* inoculated | | Non-inoculated soil | | |
| | | | | Propagules/g soil | |
	% Sporangia germination	% Damping-off		*Pythium* spp.	*P. splendens*
Hilo (conducive)	97	100	67	160	6
South Kohala (suppressive)	15	16	0	0	0

sporangial germination of *P. splendens*. It also was suppressive to damping-off of cucumber seedlings caused by this pathogen. This was in marked contrast to the Hilo conducive soil (silty clay loam, pH 5.4) (Table 2).

Pythium spp in Natural Conducive and Suppressive Soils

To study the behavior of *Pythium* spp. in these soils under natural conditions, 10 soil samples were randomly collected from sites at least 5 m apart in both South Kohala and Hilo. Six cucumber seeds were placed in soil in each of three plastic pots for each soil sample. Two weeks after planting, damping-off of cucumber seedlings occurred in Hilo conducive soil in 20 of the 30 pots (Table 2). *Pythium splendens* was detected in all tissues of diseased seedlings plated on a medium selective for pythiaceous fungi (20). Only two of the 10 soil samples collected from Hilo were free of the pathogen. None of the cucumber seedlings growing in South Kohala suppressive soil developed damping-off.

The population of *Pythium* spp. in Hilo conducive soil was 160 propagules per g of soil, of which about 4% were *P. splendens* (20,22) (Table 2). No *Pythium* was detected in South Kohala suppressive soil; however, it is not known if the suppressive effect of soil is the only reason for the absence of *Pythium* in South Kohala.

Inhibition Spectrum of Suppressive Soil

The South Kohala soil also was suppressive to damping-off of cucumber seedlings caused by *Phytophthora palmivora* (Butler) Butler and damping-off of tomato seedlings caused by *Phytophthora capsici* Leonian (15). The effect of the South Kohala suppressive soil on germination and growth of various microorganisms ranged from inhibitory to stimulatory compared with that of Hilo conducive soil (15). The suppressive soil was strongly inhibitory to germination of spores of *P. ultimum, Mucor hiemalis* Wehmer and *Fusarium oxysporum* Schlect f. sp. *cubense* (E. F. Smith) Snyder & Hansen; and moderately inhibitory to *P. aphanidermatum, Mucor ramannianus* Möller and *F. oxysporum* Schlecht f. sp. *lycopersici* (Sacc). Snyder & Hansen. It was not inhibitory to *P. palmivora, P. capsici, Neurospora tetrasperma*, Shear & B. O. Dodge, *Calonectria crotalariae* (Loose) Bell & Sobers and *Alternaria alternata* (Fries) Keissler. Growth of *Sclerotium rolfsii*, Sacc. *Xanthomonas campestris* (Pammel) Dawson, *Agrobacterium radiobacter* (Beijerinck & Von Delden) Conn, *Streptomyces alboniger* Porter et al. and *Streptomyces scabies*

(Thaxter) Waksman and Henrici was stimulated in suppressive soil; but growth of *Rhizoctonia solani* (Kühn), *Pseudomanas solanacearum* Smith and *Nocadia erythropolis* (Gary & Thornton) Waksman & Henrici was unaffected.

Transfer of Germination Inhibition from Suppressive to Conducive Soil

Suppressiveness to various pathogens has been induced in conducive soil by mixing with suppressive soil (33). To determine if suppressiveness to *Pythium* was transferable, conducive soil was mixed with 10, 25, 50 and 75% of suppressive soil and sporangial germination of *P. splendens* was determined in these soils at 0, 14, and 28 days. Inhibition of sporangial germination increased with increasing amounts of suppressive soil; however, germination inhibition in conducive soil mixed with suppressive soil did not increase during the incubation periods (15). If the source of the suppressive factor is biological, it should be able to multiply successfully in the new environment. The suppressive factor in the South Kohala soil is, therefore, considered non-biological. The transferability of the suppressive factors of potato scab-suppressive soil and *Fusarium* wilt-suppressive soil claimed by Menzies (31) and Alabouvett *et al.* (1), respectively, remain to be determined because neither report showed whether the suppressive effect of conducive soil increased after mixing with different proportions of suppressive soils.

Nature of Inhibition of Suppressive Soil

Inhibition of sporangial germination of *P. splendens* in South Kohala soil is not due to high soil pH because the soil remained suppressive after its pH was adjusted from 6.8 to 5.4. Also, conducive soil remained conducive after its pH was adjusted from 5.4 to 6.8. The inhibitory effect could not be attributed to the presence of inhibitory substances because neither volatile nor non-volatile inhibitors were detected in suppressive soil. The following evidence suggests that a combination unknown abiotic factors and a large microbial population are responsible for the suppressive effect of South Kohala soil (15):

1. Inhibition of *P. splendens* by South Kohala soil decreased with increasing depth of soil which was associated with decreasing populations of microorganisms. The inhibitory effect also was partially nullified by amendment of suppressive soil with microbial inhibitors such as streptomycin and benomyl; however, adjustment of the

total microbial population in conducive soil to the same level as that in suppressive soil did not render the former soil inhibitory.

2. Sterilization eliminated the inhibitory effect of suppressive soil. Inhibition was restored in sterilized suppressive soil by reinfestation with microorganisms from either suppressive or conducive soil, but infestation of sterilized conducive soil with microorganisms from suppressive soil did not make it inhibitory.

Calcium as the Unknown Abiotic Factor

The following lines of evidence suggest that Ca is the unknown abiotic factor and that a combination of high Ca content and a large microbial population is the cause of suppression of *P. splendens* in South Kohala soil (16):

1. Inhibition of sporangial germination of *P. splendens* in seven different types of soil whose Ca content and microbial population were known occurred only in those soils high in both Ca content and total microbial population.
2. Conducive soils low in both Ca content and microbial population became suppressive after the addition of $CaCO_3$, to increase the Ca content, and alfalfa meal to increase the population of indigenous microorganisms.
3. Conducive soils which were high in Ca but low in microbial population were converted to suppressive soils by increasing the microbial population.

Comparison of Inhibition Characteristics of Natural and Artificially Created Suppressive Soils

Establishment of the inhibition mechanism of a given suppressive soil ultimately requires confirmation whereby a conducive soil is made suppressive. Moreover, the inhibitory effect of the artificially created suppressive soil must be exactly the same as that operating naturally. This major step depends on a careful comparison of the inhibition characteristics of natural and created suppressive soil(19). Characteristics of the suppressive soil created by amending Hilo conducive soil with $CaCO_3$ and alfalfa meal were similar to those of the natural suppressive soil from South Kohala (16) and include:

1. *Association with microbial activity.* Sterilization by autoclaving or gamma irradiation nullified the inhibitory effect, and reinfestation of sterilized soil with soil

microorganisms restored the inhibitory effect in both cases.

2. *Spectrum of pathogen inhibition.* Both soils were inhibitory or stimulatory to the same species of microorganisms.

3. *Spectrum of disease suppression.* Both soils were suppressive or conducive to the same plant diseases.

4. *Response to selective inhibitors.* Streptomycin, rose bengal and benomyl reduced the inhibitory effect in both soils.

5. *Response to nutrient amendment.* The inhibitory effect of both soils was counter-acted by nutrients at similar dosage response curves.

6. *Effect of separation by a membrane.* Inhibition of sporangial germination on both soils was decreased when sporangia were separated from the soil by a polycarbonate membrane.

7. *Involvement of inhibitory substances.* Neither volatile nor non-volatile inhibitors were detected in either soil.

8. *Mode of inhibition.* Both soils were fungistatic.

9. *Transferability.* Suppressiveness was not transferable in either soil.

Similarities in inhibition characteristics suggest that the suppressive soil created artificially operates by the same mechanism as that in the natural suppressive soil and supports the hypothesis that a combination of high Ca content and high microbial population is responsible for the suppression of *P. splendens* in South Kohala soil.

ROLE OF CALCIUM IN REDUCING PYTHIUM DISEASES

Effects of Calcium on Pathogens

Few oospores of *P. aphanidermatum* and sporangia of *P. ultimum* germinated after being buried for one week in soil amended with CaO (28). The addition of CaO to soil also reduced the indigenous populations of both fungi to non-detectable levels in three weeks. When CaO-amended soil was placed in sealed jars, NH_3 which evolved from the soils reduced the germination of oospores of *P. aphanidermatum* and *P. ultimum.* Lewis and Lumsden (28), therefore, suggested that suppression of damping-off of peas on CaO-amended soil was caused by the fungicidal action of NH_3 on the pathogens; however, they did not determine whether the amount of NH_3 evolved from the CaO-amended soils was sufficient to cause the fungicidal effect under greenhouse or field conditions, however; Ca did not affect the saprophytic activity or population of *P. ultimum* in this soil (28).

The addition of Ca to soil significantly decreased sporangial germination and germ tube length of *P. splendens;* however, this also increased the concentration of root extract required for 50% germination from 0.6 to 1.7% (Table 3). Reduced germination in this case may have resulted from nutrient deprivation created by enhanced microbial activity in soil. This also may account for the reduction in diffusion distance of root exudates in Ca-enriched soil which are required for sporangial germination of *P. splendens.*

Table 3. Comparison of behavior of *Pythium splendens* on soils with and without calcium.

Behavior studied	Non-amended soil	Ca-amended (0.6%)soil
Sporangial germination[a]	90%A[d]	75%B
Average length of germ tubes [a]	460 µm A	162 µm B
Extract for 50% sporangia germination[b]	0.6%	1.7%
Sporangia germination on soil [c]	7% A	0% B

[a] Sporangia were suspended in 5% cucumber root extract before being incubated on soil. Germ tubers were measured after 6 hrs.

[b] Sporangia were suspended in different concentrations of cucumber root extract before incubation on soil. Data were obtained from dosage-response curves.

[c] Sporangia were suspended in water, placed on soil and readings made 5-6 mm from cucumber root.

[d] Values followed by the same letter for each subject compared were not significantly different at P = 0.05 according to Student's t-test.

Effect of Calcium on Hosts

After liming steamed soil infested with *P. ultimum,* reduced seedling blight of poppies but not peas indicated that Ca affected each respective host rather than the common pathogen (2); however, the observed phenomenon is probably an artifact because, in natural soils, Ca amendment also reduced seedling blight of peas caused by the same pathogen (28).

Amendment of soil with 0.6% Ca decreased damping-off of cucumber seedlings caused by *P. splendens* by about 50% (17), whereas the same treatment decreased the germination of *P. splendens* by only 15% (Table 3). This suggests the possibility that, in addition to suppression of the pathogen, Ca also may increase tolerance of the hosts to the pathogen. Amendment of soil with Ca stimulated growth of cucumber seedlings (Table 4). Some *Pythium* spp. have been classified by Garrett (10) as unspecialized parasites which are very destructive to juvenile tissues of seedlings but are restricted by mature tissues. Growth stimulation of cucumber seedlings by Ca may shorten their susceptible period and thus reduce the number of roots infected (7,27,30). Addition of Ca to soil also increased root production (Table 4) which may compensate for damage caused by the pathogen.

To determine the effect of Ca on the susceptibility of host tissues to the pathogen, cucumber seedlings grown for 5 days in Ca-amended and non-amended soils were gently washed free of soil particles with water (17). A seedling was laid between two moist filter papers placed on a large plastic petri plate. The four roots selected for inoculation were placed on filter papers with two roots on each side, and all other roots were laid in the middle of the plate and covered with soil used to grow the seedling (Fig.1). Thirty sporangia of *P. splendens* were transferred to a small paper disc (6 mm) with a microliter pipette and the inoculated side of the disc was then placed on two root tips.

Table 4. Comparison of cucumber seedlings grown in soils with and without calcium.

Subject studied	Non-amended soil	Ca-amended (0.6%) soil
Height of 10 d seedlings	3.4 cm a	5.4 cm b
Dry weight of 10 d seedlings	8.7 mg a	14.3 mg b
Average length of roots on 10 d seedlings	105 cm a	190 cm b
Infected roots after with sporangia of *P. splendens* after inoculation	81% a	21% b
Length of necrotic lesions after 3 d	20 mm a	8 mm b
Concentration of Ca in seedling	0.91% a	4.12% b

Values followed by the same letter were not significantly different at P = 0.05 according to Student's t-test.

Infection was observed under a stereomicroscope after a 3-day incubation period. Calcium decreased the incidence of necrosis of inoculated roots from 81 to 21% and reduced the average length of necrotic lesions developed in 3 days from 20 to 8 mm (Table 4).

The mechanism by which Ca amendment increased the resistance of cucumber tissue to *P. splendens* is unknown. Calcium is essential for the conversion of pectin to Ca pectate in plant tissues, and Ca pectate is resistant to degradation by polygalacturonase produced by pathogens such as *Pythium* (4,7,32). Calcium also inhibits the activity of polygalacturonase (6). Since cucumber seedlings grown in Ca-amended soil contained about 4.5 times more Ca in tissues than those grown in non-amended soil, it is possible that Ca reduced polygalacturonase activity to account for reduced disease (17).

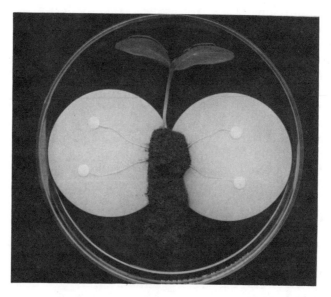

Figure 1. Method used to determine susceptibility of cucumber roots to *Pythium splendens*. Four roots from a 5-day-old cucumber seedling were inoculated each with 30 sporangia of *P. splendens* on a filter paper disc.

Multiple factors contribute to control soilborne diseases caused by *Pythium* spp. Calcium may reduce pathogen population, suppress spore germination and germ tube growth, stimulate growth of host plants, enhance root production and increase resistance of host tissues to pathogen invasion.

LITERATURE CITED

1. Alabouvette, C., Rouxel, F., and Louvet, J. 1979. Characteristics of *Fusarium* wilt-suppressive soils and prospects for their utilization in biological control. Pages 165-182 in: Soil-Borne Plant Diseases. B. Schippers and W. Gams, eds. Academic Press, London.

2.   Angell, H.R. 1950. Seedling blight. II. Soil in relation to seedling blight of opium poppy and peas. Aust. J. Agric. Res. 1:132-140.

3.   Baker, K.F., and Cook, R.J. 1974. Biological Control of Plant Pathogens. W.H. Freeman, San Francisco. 433pp.

4.   Bateman, D.F., and Lumsden, R.D. 1965. Relation of calcium content and nature of the pectic substance in bean hycoctyls of different ages to susceptibility to an isolate of *Rhizoctonia solani*. Phytopathology 55:734-738.

5.   Cook, R.J., and Baker, F.K. 1983. The Nature and Practice of Biological Control of Plant Pathogens. Am. Phytopathol.Soc., St. Paul, Minnesota. 539pp.

6.   Corden, M.E. 1965. Influence of calcium nutrient on *Fusarium* wilt of tomato and polygalacturonase activity. Phytopathology 55:222-224.

7.   Endo, R.M., and Colt, W.M. 1974. Anatomy, cytology and physiology of infection by *Pythium*. Proc. Am. Phytopathol. Soc. 1:215-223.

8.   Filonow, A.B., and Lockwood, J.L. 1979. Conidial exudation by *Cochliobolus victoriae* on soils in relation to soil mycostases. Pages 107-119 in Soil-Borne Plant Diseases. B. Schippers and W. Gams, eds. Academic Press, London.

9.   Garren, K.H. 1964. Recent developments in research on peanut pod rot. Proc. 3rd Internat. Peanut Res. Conf., Auburn, Alabama: 20-27.

10.   Garrett, S.D. 1970. Pathogenic Root Infecting Fungi. Cambridge University Press, London. 294pp.

11.   Gill, D.L. 1972. Effect of gypsum and dolomite on *Pythium* diseases of seedlings. Amer. Soc. Hort. Sci. 97:467-471.

12.   Hallock, D.L., and Garren, H.H. 1968. Pod breakdown, yield, and grade of Virginia type peanuts as affected by Ca, Mg, and K sulfates. Agron. J. 60:253-257.

13.   Hendrix, F.F., Jr., and Campbell, W.A. 1973. Pythiums as plant pathogens. Annu. Rev. Phytopathol. 11:77-98.

14.   Hora, T.S., and Baker, R. 1970. Volatile factor in soil fungistasis. Nature 225:1071-1072.

15.   Kao, C.W., and Ko, W.H. 1983. Nature of suppression of *Pythium splendens* in a pasture soil in South Kohala, Hawaii. Phytopathology 73:1284-1289.

16.   Kao, C.W., and Ko, W.H. 1986. Suppression of *Pythium splendens* in a Hawaiian soil by calcium and microorganisms. Phytopathology 76:215-220.

17.   Kao, C.W., and Ko, W.H. 1986. The role of calcium

and microorganisms in suppression of cucumber damping-off caused by *Pythium splendens* in a Hawaiian soil. Phytopathology. 61:437-438.

18. Ko, W.H. 1971. Direct observation of fungal activities on soil. Phytopathology 61:437-438.

19. Ko, W.H. 1985. Natural suppression of soilborne plant diseases. Plant Prot. Bull. 27:171-178.

20. Ko, W.H., Chang, H.S., and Su, H.J. 1978. Isolates of *Phytophthora cinnamomi* from Taiwan as evidence for an Asian origin of the species. Trans. Br. Mycol. Soc. 71:496-499.

21. Ko, W.H., and Ho, W.C. 1983. Screening soils for suppressiveness to *Rhizoctonia solani* and *Pythium splendens*. Ann. Phytopathol. Soc. Japan 49:1-9.

22. Ko, W.H., and Hora, F.K. 1971. A selective medium for the quantitative determination of *Rhizoctonia solani* in soil. Phytopathology 61:707-710.

23. Ko, W.H., and Hora, F.K. 1971. Fungitoxicity in certain Hawaiian soils. Soil Sci. 112:276-279.

24. Ko, W.H., and Hora, F.K. 1972. Identification of an Al ion as a soil fungitoxin. Soil Sci. 113-42-45.

25. Ko, W.H., Hora, F.K., and Herlicska, E. 1974. Isolation and identification of a volatile fungistatic substance from alkaline soil. Phytopathology 64:1398-1400.

26. Kobayashi, N. and Ko, W.H. 1985. *Pythium splendens*-suppressive soils from different islands of Hawaii. Soil Biol. Biochem. 17:889-891.

27. Leach, L.D. 1947. Growth rates of host and pathogen as factors determining the severity of preemergence damping-off. J. Agric. Res. 75:161-179.

28. Lewis, J.A., and Lumsden, R.D. 1984. Reduction of preemergence damping-off of peas caused *Pythium ultimum* with calcium oxide. Can. J. Plant Pathol. 6:227-232.

29. Martin, F.N., and Hancock, J.G. 1986. Association of chemical and biological factors in soils suppressive to *Pythium ultimum*. Phytopathology 76:11221-1231.

30. MᶜClure, T.T., and Robbins, W.R. 1942. Resistance of cucumber seedlings to damping-off as related to age, season of year, and level of nitrogen nutrition. Bot. Gaz. 103:684-697.

31. Menzies, J.D. 1959. Occurrence and transfer of a biological factor in soil that suppresses potato scab. Phytopathology 49:648-652.

32. Reen, L. 1971. Studies on the factors influencing virulence and enzyme activity of *Pythium* spp. on potato tubers. Indian Phytopathol. 24:74-87.

33. Schneider, R.W. 1982. Suppressive soils and plant disease. Am. Phytopathol. Soc., St. Paul, MN.

34. Toussoun, T.A. 1975. *Fusarium*-suppressive soils. Pages 145-151. in Biology and Control of Soil-Borne Plant Pathogens. G. W. Bruehl, ed. Am. Phytopathol. Soc., St. Paul, Minnesota.

35. Van der Plaats-Niterink, A.J. 1981. Monograph of the genus *Pythium*. Stud. Mycol. No. 21.

36. Vanterpool, T.C. 1940. Studies on browning root rot of cereals. VI. Further contributions on the effects of various soil amendments on the incidence of the disease in wheat. Can. J. Res. Sec. C. 18:240-257.